LEVEL RED

TEACHING RESOURCES

Unit 2
Diversity of Living Things

HOLT, RINEHART AND WINSTON
Harcourt Brace & Company
Austin • New York • Orlando • Atlanta • San Francisco • Boston • Dallas • Toronto • London

To the Teacher

This booklet contains a comprehensive collection of teaching resources. You will find all of the blackline masters that you need to plan, implement, and assess this unit. Also included are worksheets that correspond directly to the SourceBook.

Choose from the following blackline masters to meet your needs and the needs of your students:

- **Home Connection** consists of a parent letter designed to get parents involved in the excitement of the *SciencePlus* method. The letter provides parents with a general idea of what you are going to cover in the unit, and it even gives you an opportunity to ask for any common household materials that you may need to accomplish the unit's activities most economically.
- **Discrepant Event Worksheets** provide demonstrations and activities that seem to challenge logic and reason. These worksheets motivate students to question their previous knowledge and to develop reasonable explanations for the discrepant phenomena.
- **Math Practice Worksheets** and **Graphing Practice Worksheets** help fine-tune math and graphing skills.
- **Theme Worksheets** encourage students to make connections among the major science disciplines.
- **Spanish Resources** include Spanish versions of the Home Connection letter, plus worksheets that outline the big ideas and principal vocabulary terms for the unit.
- **Transparency Worksheets** correspond to teaching transparencies to help you reteach, extend, or review major concepts.
- **SourceBook Activity Worksheets** reinforce content introduced in the SourceBook.
- **Resource Worksheets** consist of blackline-master versions of charts, graphs, and activities in the Pupil's Edition.
- **Exploration Worksheets** consist of blackline-master versions of Explorations in the Pupil's Edition. To help students focus on specific tasks, many of these worksheets include a goal, step-by-step instructions, and even cooperative-learning strategies. These worksheets simplify the tasks of assigning homework, allowing opportunities for make-up work, and providing lesson plans for substitute teachers.
- **Unit Activity Worksheet** consists of an activity, such as a crossword puzzle or word search, that provides a fun way for students to review vocabulary and main concepts.
- **Review Worksheets** consist of blackline-master versions of the review materials in the Pupil's Edition, including Challenge Your Thinking, Making Connections, and SourceBook Unit CheckUp.
- **Chapter Assessments** and **End-of-Unit Assessments** provide additional assessment questions. Each assessment worksheet includes two or more Challenge questions that encourage students to synthesize the main concepts of the chapter or unit and apply them in their own lives.
- **Activity Assessments** are activity-based assessment worksheets that allow you to evaluate students' ability to solve problems using the tools, equipment, and techniques of science.
- **Self-Evaluation of Achievement** gives you an easy method of monitoring student progress by allowing students to evaluate themselves.
- **SourceBook Assessment** is an easy-to-grade test consisting of multiple-choice, true-false, and short-answer questions.

For your convenience, an **Answer Key** is available in the back of this booklet. The key includes reduced versions of all applicable worksheets, with answers included on each page.

Acknowledgments: See page 87, which is an extension of this copyright page.

Printed in the United States of America

ISBN 0-03-095675-7

2 3 4 5 6 7 8 9 021 99 98 97 96

Unit 2: Diversity of Living Things

Contents

▼ *A corresponding transparency is available. See the Teaching Transparencies Cross-Reference on the next page.*

Contents, continued

Teaching Transparencies Cross-Reference

Dear Parent,

In the next few weeks, your son or daughter will explore the diversity of living things and the reasons for this diversity. He or she will also study how different plants and animals have adapted to their particular habitats and niches. The classification of living organisms will also be examined. By the time the students have finished Unit 2, they should be able to answer the following questions to grasp the "big ideas" of the unit.

1. What is diversity? (Ch. 4)
2. Why is there diversity among living things? (Ch. 4)
3. What are some adaptations animals have for protection? for obtaining food? for attracting a mate? for locomotion? (Ch. 5)
4. How are plants adapted to survive and reproduce? (Ch. 5)
5. What is meant by natural selection? (Ch. 5)
6. What evidence did Darwin use to develop his theory of how life evolved? (Ch. 5)
7. What kinds of conditions cause species to become extinct? (Ch. 5)
8. Why do we classify things? (Ch. 6)
9. How, in a general way, are living things classified? (Ch. 6)
10. How would you classify yourself, according to Linnaeus's system? (Ch. 6)
11. What is the scientific name of your species? (Ch. 6)

You may want to ask your son or daughter some of these questions as we progress through the unit. This will give your child valuable practice communicating his or her knowledge. Don't worry if the answers do not sound like they were memorized from a textbook. With this program, students discover the answers for themselves either individually or in groups. That's both the challenge and the fun of science.

Sincerely,

The items listed below are materials that we will use in class for the various science explorations of Unit 2. Your contribution of materials would be very much appreciated. I have checked certain items below. If you have these items and are willing to donate them, please send them to the school with your son or daughter by

____.

- ○ adhesive tape
- ○ cardboard or heavy paper
- ○ craft sticks
- ○ dice
- ○ field guides for plant identification
- ○ green plants
- ○ index cards
- ○ logs (small)
- ○ magnifying glasses
- ○ markers
- ○ measuring tape, metric
- ○ plastic wrap
- ○ soft gravel
- ○ soil
- ○ spoons
- ○ sticks
- ○ string (heavy)
- ○ yarn

Thank you in advance for your help.

EXPLORATION 1

Looking for Diversity, page 79

Cooperative Learning Activity		Safety Alert!
Group size	4 to 5 students	
Group goal	to observe and record plant and animal life in a study site	
Individual responsibility	Each member of your group should have a role such as chief investigator, site preparer, materials coordinator, or data compiler.	
Individual accountability	Each group member should be able to write a short report detailing his or her findings.	

Activity 1: Diversity in a Lawn

While playing ball on the grass or stretching out in your lawn chair to enjoy the sunshine, have you ever wondered about what is happening beneath you? There may be a food-hunting expedition or a ferocious battle going on there. You can learn a lot by getting down on your hands and knees and carefully observing a small area of lawn. In this Activity you will mark off 1 sq. m of ground, preferably the day before the Exploration. Then you will study the diversity of animal and plant life in this square meter of lawn.

You Will Need

- sticks (4 per group)
- a measuring tape
- heavy string and scissors
- a magnifying glass
- your ScienceLog and a pencil
- a field guide for plant identification

What to Do

1. Find a convenient grassy area.
2. Using the measuring tape, measure out 1 m of ground and place a stick at both ends of the tape.
3. Place one end of the tape at a 90° angle from one of the sticks. Measure out 1 m, and push a stick into the ground.
4. Repeat step 3 twice to make a square.
5. Then tie string to the sticks to outline your square.
6. Observe the plant life in your study site. Fill in the table on the next page.
7. Then analyze the information you collected by considering the questions following the table.

Name ________________________________ Date ____________ Class ____________

Exploration 1 Worksheet, continued

Quick sketch of each plant	Number of same kinds (species) of plant	Description of each plant (including size, appearance, color, and any other features observed)	Common name

Questions

1. How many different types of plants did you find?

   ____

2. Which is the smallest plant you found? the biggest?

   ____

3. Which plant did you find in the greatest numbers? the fewest?

   ____

4. Do all the plants you found have some feature in common? If so, what is it?

   ____

5. How does the plant life affect the animal life in your study site?

____

____

____

____

The next day, carefully approach your staked area to see whether any larger animals, such as butterflies, are present. If you come to the area too quickly or noisily, you may scare away some animals. Using a magnifying glass, look closely for small animals that might be hidden among the plants. Fill in the table below. When you have completed the table, analyze this information by considering the questions on the next page. Then share your findings with your classmates.

Quick sketch of each animal	Number of same kinds (species) of animals	Description of each animal (including size, appearance, color, and any other features observed)	Common name

Name ______________________ Date __________ Class __________

Further Questions

1. How many different types of animals did you find?

____

2. Did you hear animal sounds but not actually see the animal?

____

3. Did you find any evidence of animal life but not see the animals themselves?

____

4. Which is the smallest animal you found? the biggest?

____

5. Which animal did you find in the greatest numbers? the fewest?

____

6. How does animal life affect plant life in your study site?

____

____

____

____

Activity 2: Diversity Around You

What to Do 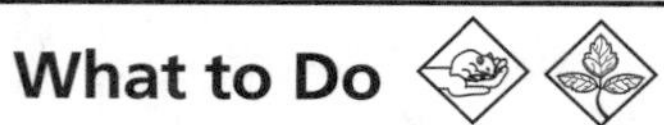

Walk around your school, your neighborhood, or a park, and fill in the tables on the next two pages. Then answer the questions that follow.

Illustration also on pages 80 and 81 of your textbook

Name ______________________ Date __________ Class __________

Exploration 1 Worksheet, continued

Quick sketch of each plant	Number of same kinds (species) of plant	Description of each plant (including size, appearance, color, and any other features observed)	Common name

Questions

1. How many different types of plants did you find?

____

2. Which is the smallest plant you found? the biggest?

____

3. Which plant did you find in the greatest numbers? the fewest?

____

4. Do all the plants you found have some feature in common? If so, what is it?

____

5. How does the plant life affect the animal life in your study site?

____

____

____

Name ______________________ Date ____________ Class ____________

Exploration 1 Worksheet, continued

Quick sketch of each animal	Number of same kinds (species) of animals	Description of each animal (including size, appearance, color, and any other features observed)	Common name

Further Questions

1. How many different types of animals did you find?

____

2. Did you hear animal sounds but not actually see the animal?

____

3. Did you find any evidence of animal life but not see the animals themselves?

____

4. Which is the smallest animal you found? the biggest?

____

5. Which animal did you find in the greatest numbers? the fewest?

____

6. How does animal life affect plant life in your study site?

____

____

____

Name
Date
Class
EXPLORATION 2
Chapter 4
Exploration Worksheet

The Puzzling Diversity of Living Things, page 82

Your goal	to learn to identify organisms by looking closely at their specific characteristics

Activity 1: Who Am I?

In the following riddles, you'll find information about the structures, habits, and habitats of some organisms. Read each riddle carefully, and think about what it tells you. Which living thing does each riddle describe? If you're having trouble figuring out what a riddle is describing, look at the pictures on this page and the next for some hints. (The pictures are not in order.) If you are really stuck, the answers are in code following the riddles. All you have to do is break the code!

Brown bear

Rainbow trout

Onion

Apple

Woodpecker

Photos also on page 82 of your textbook

Riddle 1 ______________________________

I move slowly when I am young but very quickly when I'm an adult. I eat flying insects, which I hunt near water. I have to be a strong flier to catch my food. When I stretch my four wings, I look like a helicopter. I have two more legs than a dog, and I have very large eyes. I am coldblooded and have an external (outside) skeleton. Sometimes I'm very colorful. Who am I?

Riddle 2 ______________________________

I can walk, run, and swim. I can see well, but my sense of smell is not very sharp. I am warmblooded. I am very adaptable and can live in many different environments. I really enjoy changing my environment. I care for my young for many years. I stand upright. Who am I?

Riddle 3 ______________________________

I must live in damp or wet places, avoiding the dry heat of summer and the cold of winter. If I am living in a cold climate, I become dormant in the winter. If I'm a female, I produce young by laying eggs in water. I survive by eating any moving thing that I can swallow. I can sing very well. Some of my close relatives can secrete a sticky, white poison that can kill or paralyze dogs or other enemies who may try to eat them. Who am I?

Riddle 4 ______________________________

I live in lakes, marshes, salt bays, and on beaches. I eat mostly fish and crustaceans. Although I can fly, I catch fish only by swimming. My great throat pouch is handy for scooping up fish. I fly by alternating several flaps of my wings with a glide. I always fly with my head hunched behind my shoulders. I nest on the ground in colonies. I have a wingspan of 2.5–3.0 m. My close cousins live only by the ocean, but I can venture inland. I am happy to report that these cousins are growing in number, even though they suffered from DDT poisoning a few years ago. Who am I?

Name ______________________ Date ____________ Class ____________

White pelican

Riddle 5 ______________________

I have a very high body temperature. My feet are well adapted for grasping things. I have four toes on each foot: two point forward, and two point backward. I have stiff, spiny tail feathers that act as a prop when I hunt food. I eat tree-boring insects, ants, acorns, flying insects, berries, and sap. My home, which I make myself, is a hole in a tree. I use my bill to chisel away the wood. Who am I?

Riddle 6 ______________________

I have pointed, green stalks above the ground and a rounded, brown bulb below. People must pull me out of the soil before I can be useful to them. Cooks use me to improve the taste of food. If people bite me, I can bite back, making their eyes water. Who am I?

Toad

Riddle 7 ______________________

I live in cold, well-oxygenated water, and I'm a fast, strong swimmer. I am slim, sleek, and colorful. I'm a carnivore; I eat mostly insects and smaller members of my own kind. I spawn my eggs during the spring in small, clear streams. I'm coldblooded. Who am I?

Riddle 8 ______________________

I undergo wondrous changes during my life. At the beginning, I am a sweet-smelling, pink-and-white blossom. Later I'm a hard, green ball that makes your eyes water and your mouth pucker if you try to eat me. Finally, I become a sweet, juicy, red or yellow fruit. People say I keep physicians away. Who am I?

Elk

Riddle 9 ______________________

I am a big animal. My mass is about 225 kg, but my tail is only about 15 cm long. I am dark in color. Generally, I live on forest floors and in thickets. When it starts to get really cold, I enter my shelter for the winter. I don't have very good sight, but my senses of hearing and smell are keen. Using these senses, I find lots of food—small animals, insects, garbage, leaves, grasses, berries, nuts, and fruits. Who am I?

Riddle 10 ______________________

I am warmblooded and hairy. I feed milk to my young. I chew my cud, and I have a complex stomach. The males of my kind have huge, branching antlers. I have a heavily maned neck. Humans, wolves, and mountain lions are my only enemies, but mountain lions usually won't attack me when I'm fully grown. My young are not camouflaged from these enemies until their winter hair grows out. Sometimes you can hear the males of my kind give a high-pitched bugle call. If this call is answered by another male, a battle may follow. Who am I?

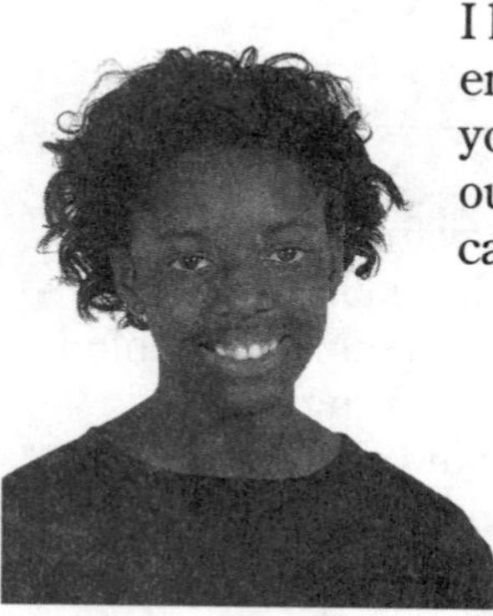
Human

Dragonfly

Photos also on page 83 of your textbook

Name ______________________ Date ____________ Class ____________

Here are the coded answers to the riddles:

1. CQZFNMEKX
2. GTLZM
3. SNZC
4. VGHSD ODKHBZM
5. VNNCODBJDQ
6. NMHNM
7. QZHMANV SQNTS
8. ZOOKD
9. AQNVM ADZQ
10. DKJ

Hint for decoding: ZMS = ANT

Activity 2: You Be the Riddler!

Read the riddles again carefully, this time noticing which characteristics of the organisms are used to describe their diversity. Then try writing your own riddles for some of the living things pictured on pages 84–85 of your textbook. Before you start writing, do some research to find out about the organisms you chose. After writing your riddles, see if your classmates can solve them.

Name ______________________ Date __________ Class __________

Challenge Your Thinking, page 86

1. Seven Days of Diversity

In your ScienceLog, write a series of log entries that describe the different plants and animals you see during one week. Where did you see them? What were they doing? Were there any changes over the week? Did you see anything that surprised you? Compare your log entries with those of your classmates.

2. Tourist Attraction

Your city's tourist bureau has asked the members of your class to write a section in their new brochure called "Diversity in My Community." What would you write about the diversity of living things in your community in order to attract tourists to visit?

3. Now You See It, Now You Don't

Create a pictorial time line that shows Krakatau before the volcano erupted and that includes the changes on Rakata up to 1985.

Name ______________________ Date ____________ Class ____________

4. Invisible Differences

Look at the three animals pictured on page 87 of your textbook. What similarities do you see? What differences do you see? Do you think there are differences and similarities that are not visible? Explain your reasoning.

5. They're Everywhere!

What plants do you sit on? What plants do you wear, eat, write on, and sleep in? Think about the ways that plants affect your life, and list as many as you can. Describe what your life would be like if there were not so much diversity in plants.

Illustration also on page 87 of your textbook

Name ______________________ Date ____________ Class ____________

Chapter 4
Assessment

Word Usage

1. Use all of the following terms in one or two sentences to show how they are related:

a. diversity, organisms, biologists, differences

____

____

____

b. appearance, habitat, living things, eating, characteristics, distinguish

____

____

____

Correction/ Completion

2. The statements below are either incorrect or incomplete. Your challenge is to make them correct and complete.

a. The diversity of life can be seen only in certain regions of the world among specific organisms.

____

____

____

b. There are more than ______________ known kinds of living things in the world, but the actual number is estimated to be between ______________ and ______________.

Short Response

3. Compare and contrast bats and birds.

____

____

____

____

____

____

Name ______________________________ Date ____________ Class ____________

CHALLENGE 4

Illustration for Interpretation

4. Look at the illustration below. How is the diversity of life influencing these two organisms?

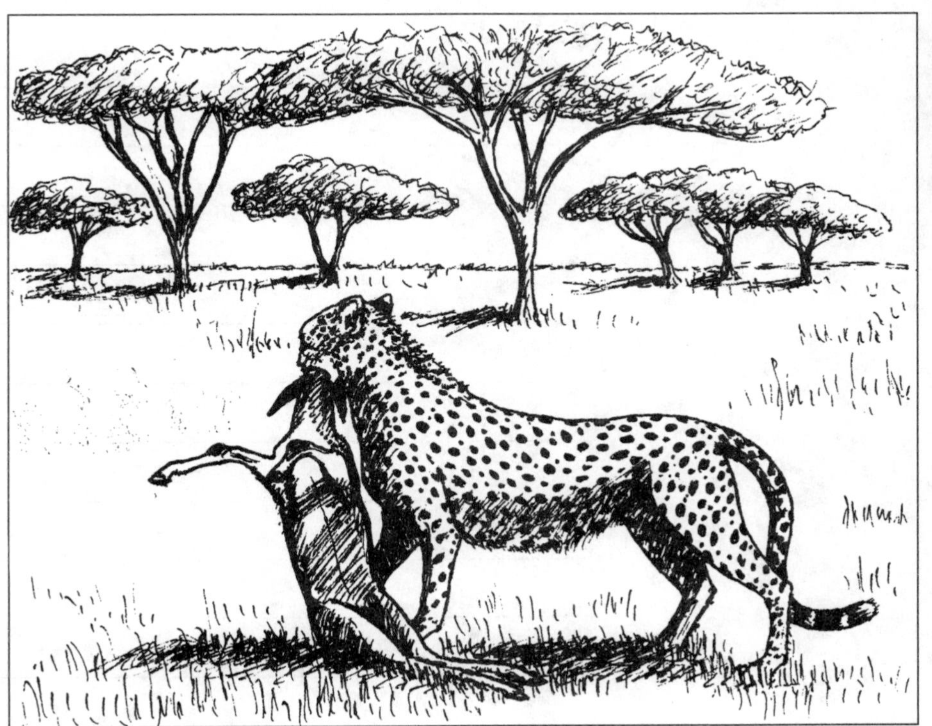

____

____

____

____

____

____

____

____

____

____

____

____

____

Name ____________________ Date __________ Class __________

CHALLENGE 2

Short Essay

5. Suppose that a layer of dust in the atmosphere caused a worldwide reduction in the average temperature. If this condition were to last for hundreds of years, how might the diversity of life be affected in the place where you live?

Name ______________________ Date __________ Class __________

The Case of the Peppered Moth, page 93

Read the following account and answer the questions on the next page.

Research Subject

Peppered moths have lived in the forests of England for thousands of years. They rest on the trunks of trees during the day and are a source of food for many birds. Peppered moths vary in color, from light-colored to dark-colored.

Research Problem

Did air pollution, which covered tree trunks with black soot from the many new factories during the Industrial Revolution of the 1800s, affect the survival of the peppered moth?

Conditions

Before the Industrial Revolution, the tree trunks were light-colored. The trunks and branches were also covered with silvery white lichens. As the Industrial Revolution progressed, pollution killed the lichens and blackened the tree bark.

Hypothesis

In the 1950s, Oxford University professor H.B.D. Kettlewell and his assistants formed a hypothesis about the peppered moth. Form your own hypothesis. It should include the effect that you think the Industrial Revolution had on the peppered moth's survival.

Procedure

With his assistants, Kettlewell performed an experiment to test his hypothesis. They used the following procedure:

1. They located two areas. One was a wooded area with lichen-covered oak trees. The other was a wooded area that had been subjected to pollution for many years.
2. They released a known number of light-colored and dark-colored peppered moths into each area.
3. After a given amount of time, they recaptured as many moths as they could.

Results

In the unpolluted area, more light-colored moths survived. In the polluted area, more dark-colored moths survived.

Name ____________________ Date __________ Class __________

Discussing the Results

1. Do these results support your hypothesis? (They supported Kettlewell's hypothesis.)

2. How can you explain the results?

3. Why didn't Kettlewell release the moths only into the polluted area?

4. Do Kettlewell's findings support the theory of natural selection? Explain your answer.

Name ______________________________ Date ____________ Class ____________

Darwin's Finches, page 94

Use the following information and table to answer the questions on the next page.

In 1835, Darwin spent five weeks visiting the Galápagos Islands as the naturalist on a ship called the H.M.S. *Beagle.* He observed, recorded, collected, and preserved everything he could of the islands' natural history. There were many strange and colorful animals, but what interested him most were drab little birds that made unmusical sounds—finches. The finches on all the islands resembled each other closely, except for one set of features—the size and shape of their beaks. Separated for thousands of years on the different islands of the Galápagos chain, the birds had adapted in unique ways to their environments. The differences among the finches of the various islands are shown in the table.

Differences Among Some of Darwin's Finches

Name of the finch	Feeding habit	Form of beak
small tree finch	Uses delicate bill to eat aphids and small berries.	
large tree finch	Grinds fruit and insects with parrotlike bill.	
small ground finch	Uses pointed bill to eat tiny seeds and to pick ticks from iguanas.	
large ground finch	Conical bill enables it to eat large, hard seeds.	
cactus finch	Long bill probes for nectar in cactus flowers.	

Illustrations also on page 94 of your textbook

Name ______________________ Date ____________ Class ____________

Questions

1. How is the structure of the beak well suited to the diet of each group of finches?

2. Do the differences in Darwin's finches support the theory of natural selection? Give reasons for your answer.

3. Scientists have speculated that Darwin's finches reached the Galápagos Islands from the mainland of South America as a single flock perhaps a million or more years ago. Think about the following questions, and explain your answers:

   a. What do you think the original finches looked like? Why?

   b. Is it possible that the original birds were various species that arrived on the islands at different times?

Name ____________________ Date __________ Class __________

c. Assume that one flock of finches gave rise to the 14 different species now existing on the islands. If this occurred, would it be significant that the Galápagos chain consists of many small islands rather than one large one?

____

____

____

____

____

d. What advantages would the finches have had in arriving on the islands under the following conditions?

i. There were no other species with exactly the same diet.

____

____

____

ii. There were no predators.

____

____

____

iii. There were no parasites to live on the finches and weaken them.

____

____

____

4. How has diversity helped Darwin's finches survive?

____

____

____

____

____

____

____

Name ______________________________ Date ____________ Class ____________

A Tasteful Activity — Teacher Demonstration

Use this exercise to introduce The Fine Art of Survival on page 97 of your textbook.

You Will Need

- toothpicks
- an onion
- a pear
- an apple
- a potato

Note: *Cut the onion, pear, apple, and potato into bite-size pieces before beginning.*

What to Do

1. Ask students: How do we know whether something tastes good? *(Expected response: We perceive taste with our tongue.)*
2. Ask for a volunteer (or assign students to groups) to find out for sure how we perceive taste.
3. Have the volunteer close his or her eyes and pinch his or her nose tightly shut.
4. Using toothpicks, pick up the sample foods and hand them, one at a time, to the volunteer.
5. Have the volunteer taste and attempt to identify each sample. The student's ability to taste should be greatly diminished by his or her inability to smell.
6. Have the volunteer release his or her nose but keep his or her eyes closed. Hold one food sample under the student's nose while he or she tastes a second sample. The student should "taste" the sample he or she can smell. This is especially effective if the sample being smelled has a stronger smell than the one being eaten.

Discussion

Use the following information to discuss the role of taste and smell in survival:

Although smell and taste are two distinct systems, most of what we call taste is really smell. Our sense of taste detects only sweet, sour, salty, and bitter; other components of flavor come from smells. The tissue responsible for the sense of smell consists of 12 million specialized cells, each with 10–20 hairlike growths called *cilia*. Each cilium has a receptor that binds an odorant molecule, which triggers a nerve impulse and sends the message to a part of the brain called the *olfactory bulb*. Taste comes from 10,000 taste buds found on the tongue, cheeks, throat, and the roof of the mouth.

The senses of taste and smell have helped humans survive and evolve. One theory suggests that humans evolved the ability to detect bitter flavors at the back of the tongue because the primary cues of toxicity are bitter taste and pungent odors. This could allow a person a final chance to spit out suspect foods. According to Darwin's theory of natural selection, humans who were able to distinguish between food and toxins by smell and taste were more likely to survive and reproduce.

Name ______________________ Date __________ Class __________

Chapter 5
Resource Worksheet

The Fine Art of Survival, page 97

Adaptations are inherited features that help increase an organism's chances of surviving and reproducing. Both animals and plants have many different kinds of structures and behaviors that help them survive in their environments. Below are three categories of adaptations to discuss: adaptations for obtaining food, adaptations for protection, and adaptations for locomotion. The organisms on pages 96 and 97 of your textbook will give you some hints.

1. Name some examples of the great variety of structures and behaviors (adaptations) that animals and plants have for obtaining their food.

2. Can you think of animal and plant adaptations that could be used for protection? Give some examples.

3. Consider an animal's locomotion—its movement from place to place. Name some adaptations that various animals have for locomotion. Is each adaptation you thought of related in some way to the organism's habitat? Explain.

Chapter 5

Name ______________________ Date ____________ Class ____________

Using the Theme of Changes Over Time

This worksheet is an extension of the theme strategy outlined on page 108 of the Annotated Teacher's Edition. It is designed to correspond to Lesson 3, The Value of Diversity, which begins on page 106 of the Pupil's Edition.

Focus question	How might human activities affect the diversity of life on Earth?

The Earth's atmosphere has changed considerably in the past 4.5 billion years. The graph below shows the changes that have occurred in the levels of gases in the atmosphere.

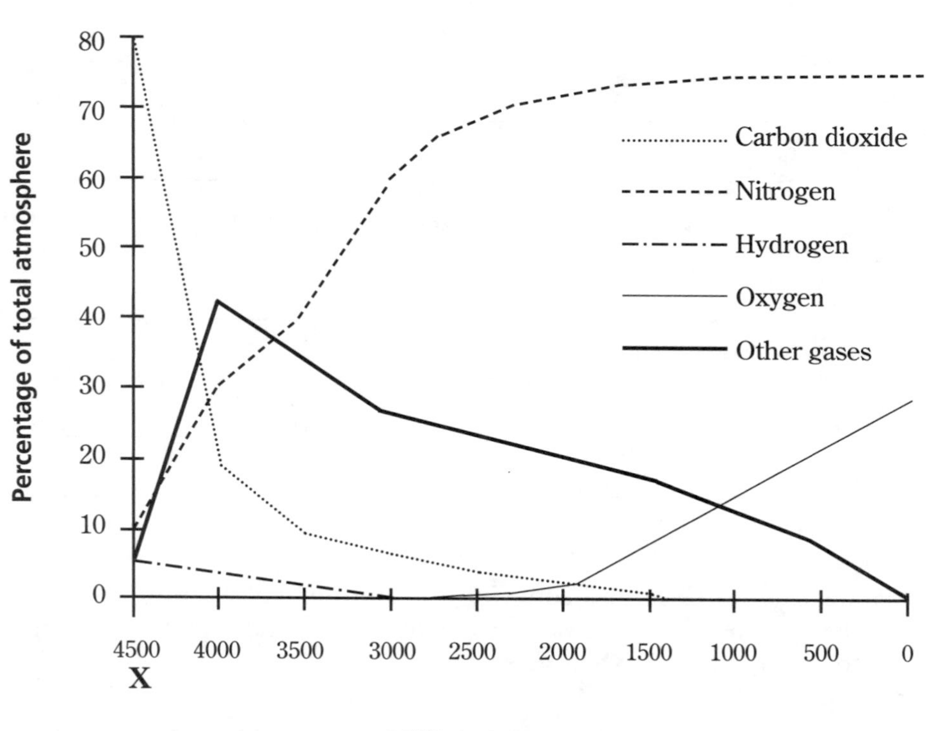

Using the data in the chart on the next page, indicate with an "X" when a significant event took place. The first event has been marked for you.

Name ______________________ Date ____________ Class ____________

Event	Millions of years ago
Formation of the Earth	4500
First living organisms evolve in the sea.	3500
Photosynthesis begins in bacteria.	3000
First multicellular organisms evolve.	630
Plants, animals, and fungi first invade land.	410
Amphibians are the first vertebrates on land.	350
The first humans appear.	2
Tropical rain forests are being destroyed, damage to ozone layer is increasing, and pollution of atmosphere is evident.	0

Questions

1. What is remarkable about the graph at the time when photosynthesis begins in bacteria and when plants first appear?

____

____

____

2. Which gas do you think is likely to increase the most during your lifetime? Why?

____

____

____

3. What effects does damaging the atmosphere have on diversity?

____

____

____

____

____

____

____

____

Name ______________________ Date ____________ Class ____________

EXPLORATION 1

Chapter 5
Exploration Worksheet

The Extinction Game, page 108

Your goal	to learn about the factors that lead to the survival or extinction of a species

Animals may become *extinct* when their habitat changes significantly. The Extinction Game will help you discover the kinds of changes that may affect the survival of certain animals. Two to eight people can play at one time.

Getting Started

Using the attached worksheet as a guide, make 20 cards each for the following animals: whooping crane, bowhead whale, wood bison, sea otter, Peary caribou, eastern cougar, white pelican, and Eskimo curlew. Have 20 blank animal cards on hand in case some of the animal populations grow beyond 20 members. Make two copies of each extinction card shown on the attached worksheet. Also make one token for each animal, to be used on the gameboard path.

1. To play the game, choose one of the animals shown on the attached worksheet.
2. Begin with a population of 20 of your chosen animal. To be a winner in the game, you need to finish before your animal becomes extinct. You will be using two sets of cards for this game: "animal cards," which represent each of the eight animals, and "extinction cards," which tell you how many members to add or subtract from your animal's population.
3. Place the extinction cards facedown in a pile on the table. (After drawing an extinction card, place it on the bottom of the pile.)
4. Roll a die to determine who begins—the highest goes first.
5. Then, in turn, each player throws the die and moves the number of spaces shown on the die. Remove animal cards from your pile or add them to your pile as required by the extinction cards. If your animal becomes extinct, you are out of the game.
6. Extinction is a process that usually happens over a long period of time. This period may be represented by traveling the path a second time with your surviving population.
7. After you've finished the game, discuss what you learned about extinction with your classmates.

Name ______________________ Date ____________ Class ____________

Animal Tokens

Animal Cards

White pelican

Peary caribou

Eastern cougar

Sea otter

Whooping crane

Wood bison

Bowhead whale

Eskimo curlew

Name ______________________________ Date ____________ Class ____________

Extinction Cards

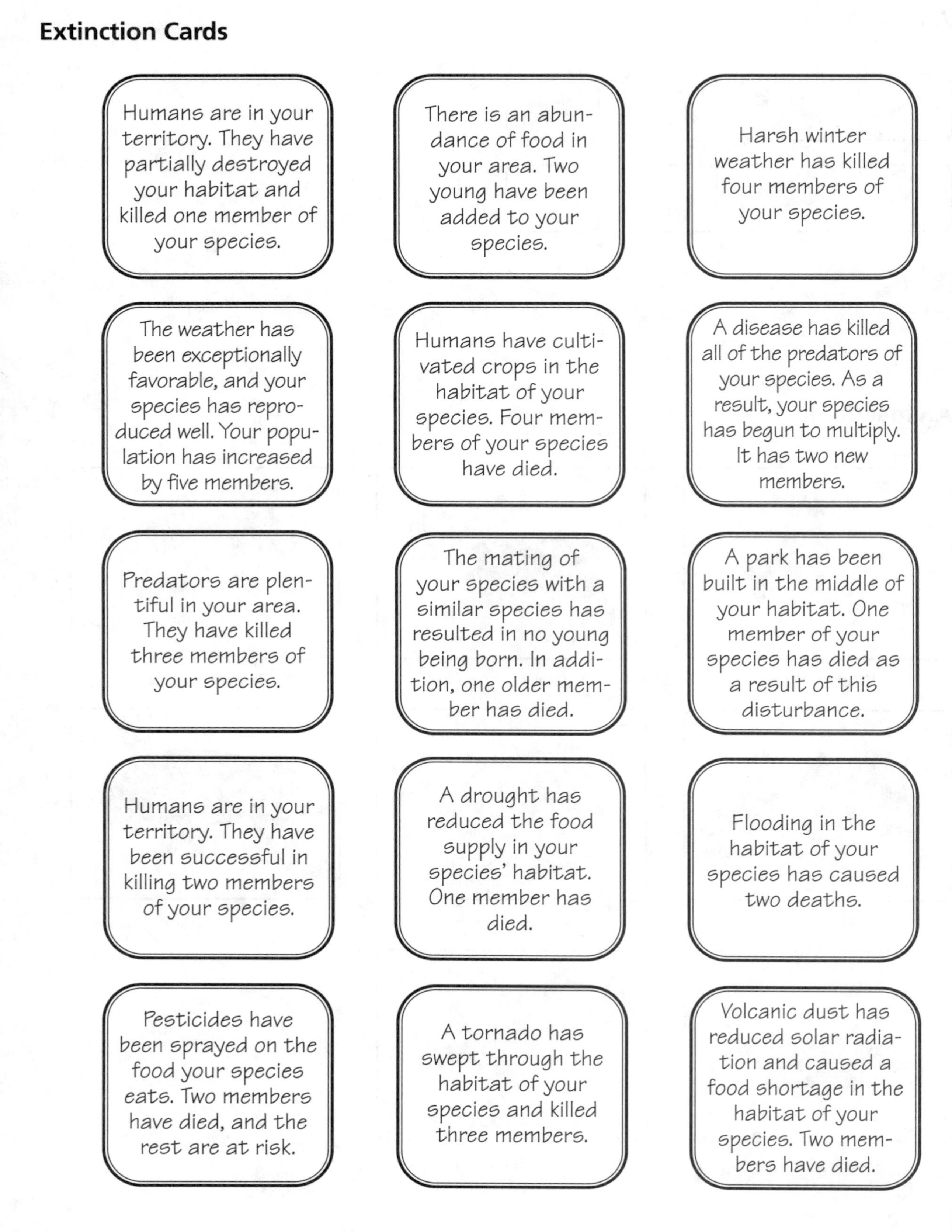

Humans are in your territory. They have partially destroyed your habitat and killed one member of your species.
There is an abundance of food in your area. Two young have been added to your species.
Harsh winter weather has killed four members of your species.
The weather has been exceptionally favorable, and your species has reproduced well. Your population has increased by five members.
Humans have cultivated crops in the habitat of your species. Four members of your species have died.
A disease has killed all of the predators of your species. As a result, your species has begun to multiply. It has two new members.
Predators are plentiful in your area. They have killed three members of your species.
The mating of your species with a similar species has resulted in no young being born. In addition, one older member has died.
A park has been built in the middle of your habitat. One member of your species has died as a result of this disturbance.
Humans are in your territory. They have been successful in killing two members of your species.
A drought has reduced the food supply in your species' habitat. One member has died.
Flooding in the habitat of your species has caused two deaths.
Pesticides have been sprayed on the food your species eats. Two members have died, and the rest are at risk.
A tornado has swept through the habitat of your species and killed three members.
Volcanic dust has reduced solar radiation and caused a food shortage in the habitat of your species. Two members have died.

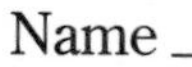
Name ______________________ Date __________ Class __________

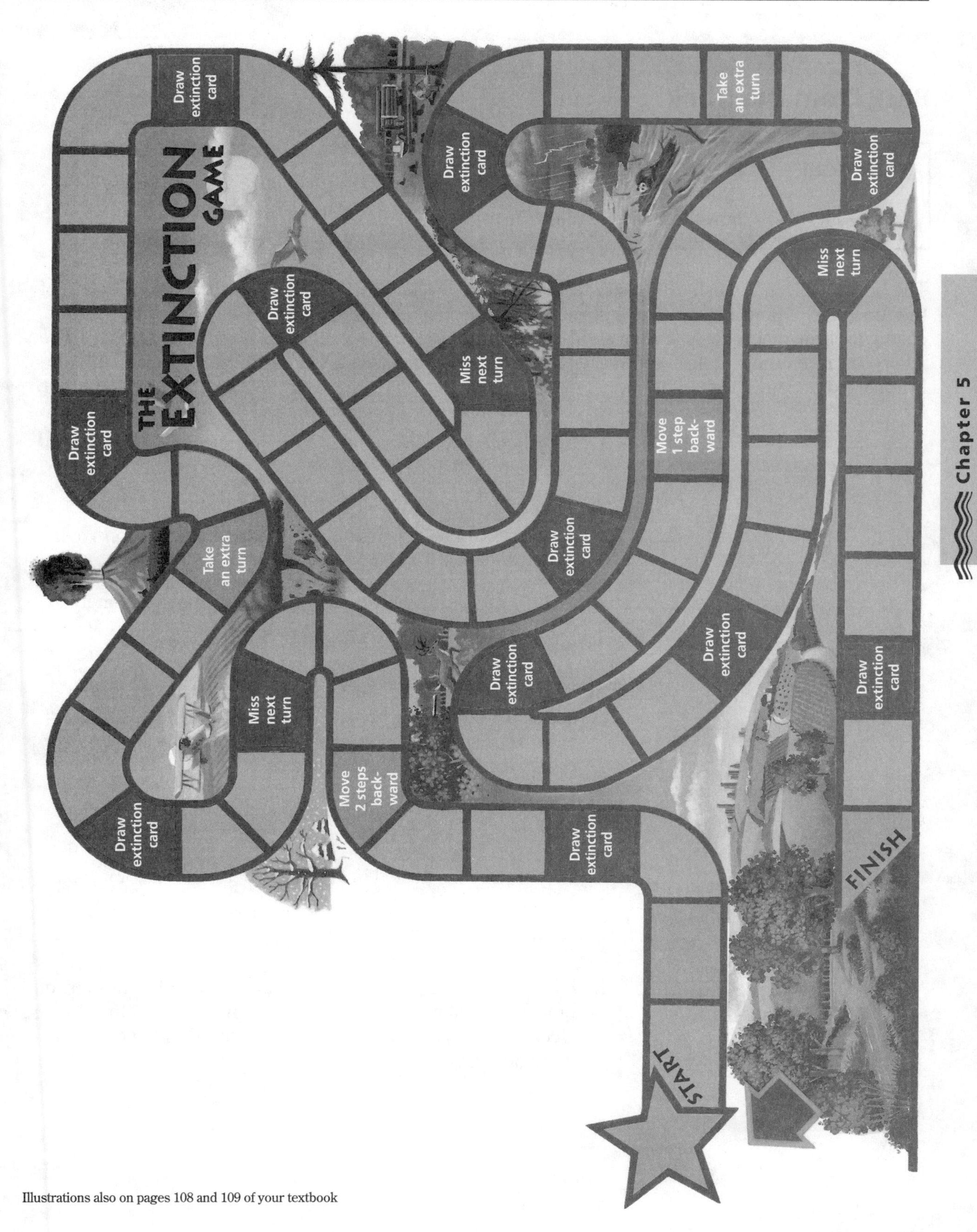

Illustrations also on pages 108 and 109 of your textbook

Name ______________________________ Date ____________ Class ____________

EXPLORATION 2

Chapter 5
Exploration Worksheet

Back From the Brink, page 110

Your goal	to learn about factors that can lead to the extinction of a species and to determine how those factors might be avoided

The brown pelican is a fish-eating coastal bird that nests along the Atlantic, Pacific, and Gulf Coasts of the Americas. Below are some facts about brown pelicans on West Anacapa Island, a major breeding colony for brown pelicans in California. The facts trace the brown pelican's history from the days when it lived in large numbers on West Anacapa Island, through its decline, and to its comeback from the brink of extinction. The facts are not in the right sequence, however. Your task is to put them in order.

______ **a.** Investigators discovered that a chemical company had been dumping DDT into the Los Angeles sewer system for some time.

______ **b.** An average of 5000 pairs of brown pelicans nested on West Anacapa Island from 1985 to 1989.

______ **c.** In 1973 the United States placed the brown pelican on its list of endangered species—those species that may not survive in the wild unless they are protected.

______ **d.** When brown pelicans ate DDT-contaminated fish, the DDT accumulated in their bodies. This caused the shells of their eggs to be so thin and fragile that they often broke while the eggs were being laid or incubated.

______ **e.** In 1970, only one brown pelican hatched on West Anacapa Island.

______ **f.** Brown pelicans nested in large numbers on West Anacapa Island.

______ **g.** In 1972 the use of DDT was banned.

______ **h.** DDT worked well in killing mosquitoes and other insects, but biologists discovered that DDT had contaminated the Pacific Ocean and the fish in it. This caused considerable harm to the food chain.

Name ______________________ Date ____________ Class ____________

Another Close Call

Examine the diagram below closely. What does it tell you about the survival of the brown pelican in Texas and Louisiana? Record your observations below.

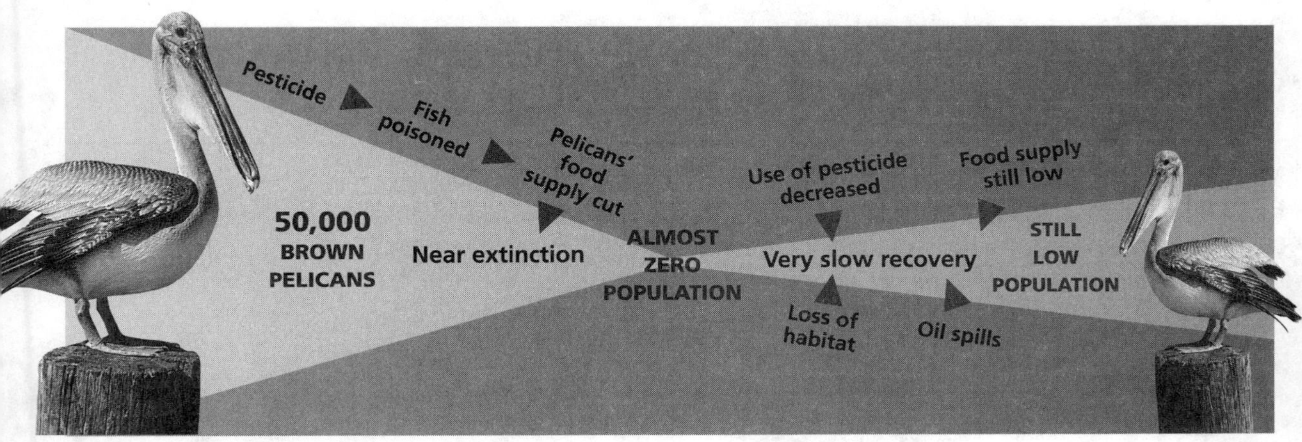

Illustration also on page 110 of your textbook

Chapter 5

Name ______________________ Date ____________ Class ____________

May the Best Animal Win

Complete this worksheet at the end of Chapter 5, which begins on page 88 of your textbook.

Marguerite and Susan both have pets. Marguerite's pet is a small house spider. She wants to race her spider against Susan's pet, a green turtle. Because Susan is sure her turtle is slower, she agrees to the race only if her turtle is given a 7 cm head start. Marguerite agrees. Susan recorded the results of three different races in the graphs below.

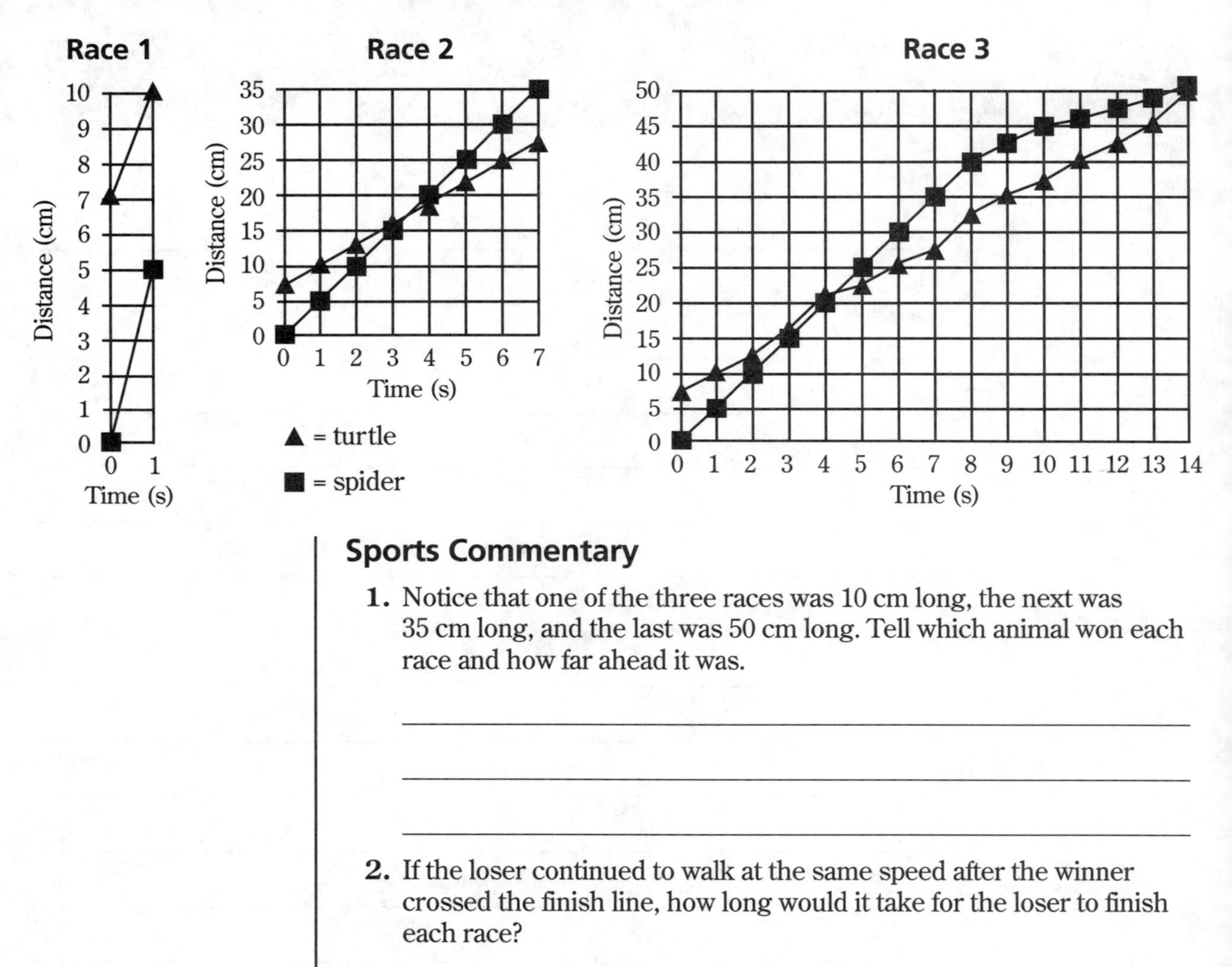

Sports Commentary

1. Notice that one of the three races was 10 cm long, the next was 35 cm long, and the last was 50 cm long. Tell which animal won each race and how far ahead it was.

____

____

____

2. If the loser continued to walk at the same speed after the winner crossed the finish line, how long would it take for the loser to finish each race?

____

____

____

____

3. How do the spider's and the turtle's walking rates differ during the third race?

4. How do these differences relate to the animals' daily activities?

5. This race was held on level, dry ground. Would the results be different if it was held in water? Why?

6. How do these races relate to the "survival of the fittest"?

Name ______________________ Date __________ Class __________

Challenge Your Thinking, page 111

1. Tall Tales

Evidence in the fossil record suggests that the ancestors of modern giraffes had very short necks. They lived on grasslands in Africa where there was a lot of vegetation that they were able to use for food. There were short grasses, bushes, and tall trees. There is also evidence of long periods of drought in that region of Africa.

How do you think Lamarck would explain the development of giraffes with long necks? How would Darwin explain it?

Name ______________________ Date __________ Class __________

2. To the Editor

An article in the local newspaper stated, "Everything that humans do to the environment causes animals to become extinct." Write a letter to the editor stating whether you agree or disagree with this statement. Give several reasons for your viewpoint.

3. A Different World

Imagine what the world would be like if

- all plants were 5 cm tall.
- all bears were black.
- all rabbits were white.
- no insects could fly.
- only seals lived in the ocean.

For one of these situations, list all of the ways you think the world would be different from the way it is now.

Illustration also on page 111 of your textbook

Name ______________________ Date __________ Class __________

4. Critter Creation

Appendages are adaptations designed to help an animal perform various specialized tasks. Your thumb is an appendage; in fact, your entire hand and arm is an appendage. Appendages help an animal live successfully in its environment.

In your ScienceLog, design and draw an animal with appendages that will give it the following characteristics:

a. The animal lives in water.

b. Its heavy body needs a lot of support as the animal walks.

c. It can walk on the bottom of a body of water for many kilometers without stopping.

d. It can dart away suddenly from its enemies by swimming.

e. It can create water currents to bring food in the water to its mouth.

f. It tests its food before eating it.

g. It has appendages that enable it to hold large pieces of food.

h. It can break apart hard bits of food.

i. Its diet includes shelled animals.

j. It has appendages that enable it to hold its young.

k. It has formidable defensive weapons.

Does the animal you drew look like any animal you have seen before? Do you think an animal exists that has all of the above characteristics? Explain your reasoning.

____

____

____

____

____

____

____

____

____

____

____

Illustration also on page 112 of your textbook

Name ______________________ Date ____________ Class ____________

Chapter 5
Assessment

Word Usage

1. Use all of the following terms in one or two sentences to show how they are related:

   a. adaptations, organisms, natural selection, reproduce, environments

   b. camouflage, mimicry, changes, snowshoe hare, avoid, hornet fly

Correction/ Completion

2. The statements below are either incorrect or incomplete. Your challenge is to make them correct and complete.

   a. Members of a species can display adaptations to a change in their environment within a short period of time.

Name ______________________________ Date ____________ Class ____________

b. ______________________________ believed that an organism could acquire a favorable characteristic during its lifetime and pass that characteristic on to its offspring.

c. ______________________________ believed that organisms with favorable characteristics were more likely to survive, reproduce, and pass those favorable characteristics to their offspring.

Short Response

3. For each adaptation on the table below, mark an "X" to show whether the adaptation helps primarily with locomotion, protection, or getting food.

Adaptation	Locomotion	Protection	Obtaining food
Eagles have extremely sharp vision.			
Some monkeys can hold branches with their tails.			
Roosters have a sharp spur on each foot.			
Ivy plants have flat, dark green leaves.			
Mesquite trees grow spikes on their branches.			
Bats have thin membranes between their fingers.			

CHALLENGE

Numerical Problem

4. Every spring, Nina begins to work on her lawn and gardens. First she counts the number of dandelions on her lawn. She records the number each year in order to determine how this particular area resists or succumbs to weeds. Two years ago, there were 18 dandelions, last year there were 23 dandelions, and this year there are 28 dandelions.

a. If the dandelions keep growing at the same rate, how many dandelions will there be next year?

____

____

b. Nina is frustrated by the greater amount of weeding she must do every year. Explain why the dandelions are so successful at reproducing.

CHALLENGE 2

Short Essay

5. Applying the theory of natural selection, explain how the ancestors of the air-breathing porpoise, who were land animals, developed the ability to live in the water.

Name ______________________ Date ____________ Class ____________

Classification Pyramid Teacher's Notes

This worksheet corresponds to Transparency 16 in the Teaching Transparencies binder.

Suggested Uses

Use as a visual aid with the topic listed below.

Grouping Living Things, page 114

Use the transparency to extend the study of the classification of living things.

This transparency shows the complete classification of the mountain lion, *Felis concolor.* The pyramid can be revealed to the students one layer at a time, starting at the bottom, by using an opaque sheet of paper as a cover. Discuss each level of classification.

Use the transparency with the transparency worksheet on the next page for reteaching or review. Please note: The worksheet is a reproduction of the actual transparency with certain labels omitted. For answers to that worksheet, see the transparency.

Possible Extension Questions

1. How are living things classified?
2. Arrange the following classification groups in proper sequence: kingdom, order, species, phylum, family, class, genus.
3. What distinguishes an animal as a member of the class Mammalia? of the order Carnivora?

Answers to Extension Questions

1. All living things are divided into groups according to certain basic similarities. Then each of these groups is divided, subdivided, and so on, until the level of individual species is reached.
2. Kingdom, phylum, class, order, family, genus, species
3. Mammals are animals that have fur and nurse their young. Carnivores are animals that have fangs and eat flesh. (Students may wish to research other characteristics that biologists use for classification.)

Name ______________________ Date ____________ Class ____________

Classification Pyramid

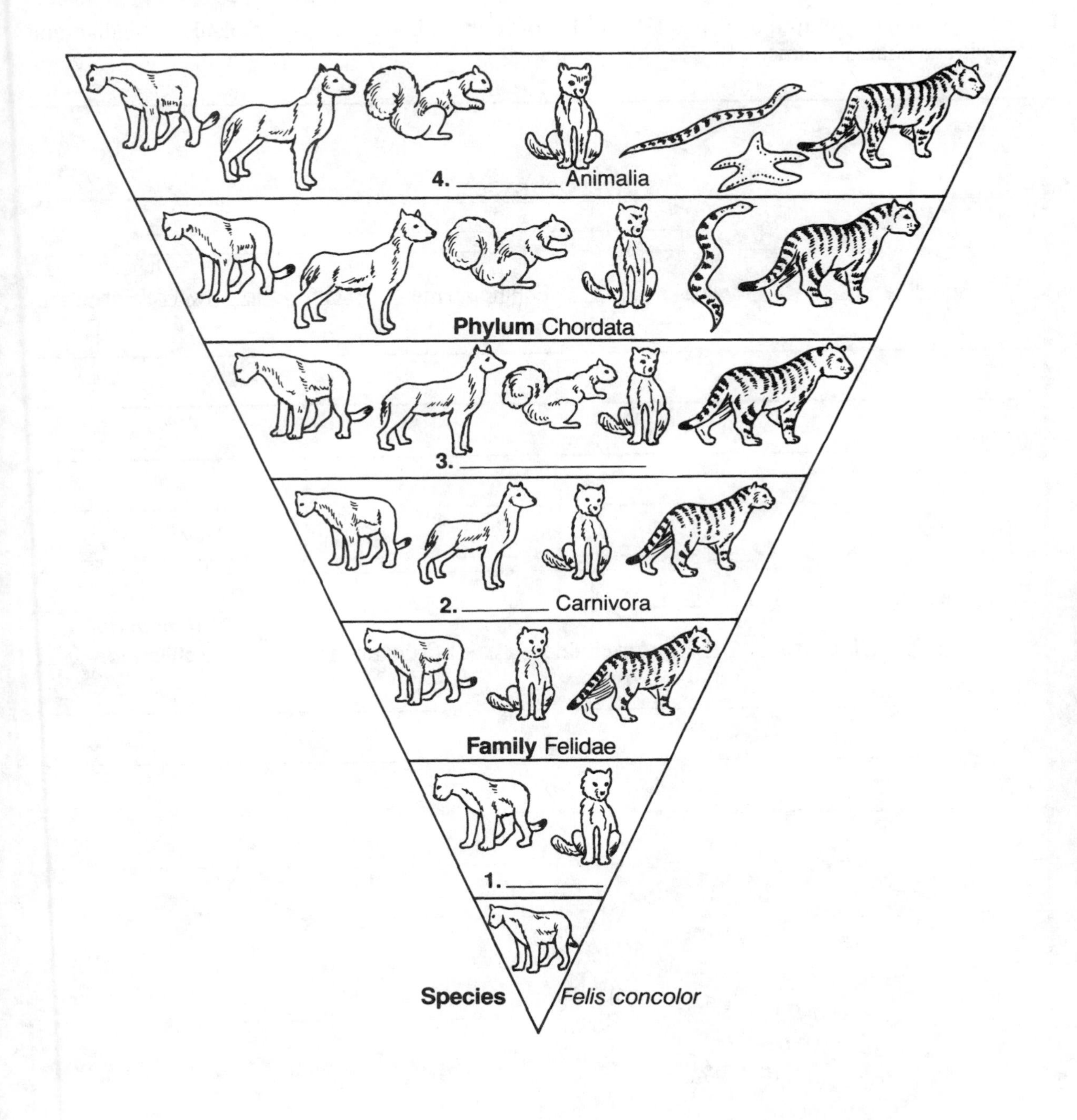

Chapter 6

Name ______________________ Date ____________ Class ____________

A Simplified Classification System for Invertebrates, page 123

Look at the animals pictured on pages 120 and 121 of your textbook. Fill in each blank in the diagram below with an animal's name.

Invertebrates

Annelids	Mollusks	Echinoderms	Arthropods	Other invertebrates*
______	______	______		______
______	______	______		______
	______	______		______
	______			______

Arthropods

Insects	Arachnids	Crustaceans	Diplopods/ chilopods
______	______	______	______
______	______	______	______
______	______		

*"Other invertebrates" includes other worms, such as flatworms and unsegmented roundworms, along with many water animals, such as jellyfish, sea anemones, corals, and sponges.

Questions

Now think some more about the subgroups.

1. Did you have any problems deciding which subgroup each invertebrate belonged to? If so, which one(s)?

2. It's interesting to think about where invertebrates live. How many are found in water?

   in moist places?

   on dry land?

3. Does it appear that the structure of invertebrates enables them to live successfully in various places? Why do you think this is the case?

Name ______________________ Date __________ Class __________

Chapter 6
Resource Worksheet

Putting It All Together, page 126

Many animals are shown on pages 126–129 of your textbook. Your task is to classify them using the classification tables below. These tables are an overall classification system for animals. They bring together everything you have learned about classification for the animal kingdom.

If your classroom contains any living or preserved specimens that are not represented in the pictures, classify them as well.

Animal Kingdom Classification System

Invertebrate subgroups		Examples
Annelids		
Mollusks		
Echinoderms		
Arthropods	Insects	
	Crustaceans	
	Arachnids	
	Diplopods/chilopods	
Other invertebrates		

Vertebrate subgroups	Examples
Fishes	
Amphibians	
Reptiles	
Birds	
Mammals	

Name ______________________ Date ____________ Class ____________

EXPLORATION 1

Chapter 6
Exploration Worksheet

Tracing Similarities and Differences, page 134

Cooperative Learning Activity

Group size 4 students

Group goal to study the presence of inherited characteristics in a population

Individual responsibility Each member of your group should have a role such as group leader, data manager, or primary investigator.

Individual accountability Each group member should complete his or her own data table and use it to write a brief summary of the investigation.

What to Do

1. In the classroom, work in groups of four.
2. Examine six inherited characteristics for each member of the group.
3. Record each member's characteristics in the table below.
4. Repeat this procedure at home with family members or other relatives.

Characteristic		Student				Family	
		You	2	3	4	1	2
Hand folding, thumb position	1. Left over right						
	2. Right over left						
Ear lobes	3. Attached						
	4. Free						
Hairline	5. Pointed						
	6. Straight						
Tongue rolling	7. Can roll						
	8. Can't roll						
Middle segment of fingers	9. Hair						
	10. No hair						
Toe next to big toe	11. Same length or longer than big toe						
	12. Shorter than big toe						

Chapter 6

Analyze Your Data

1. Study your table and compare your results with those of the other groups in your class.

2. Calculate the percentage of individuals in your group who have each characteristic. What does this tell you?

3. Does anyone have all of characteristics 1, 3, 5, 7, 9, and 11? Does anyone have all of characteristics 2, 4, 6, 8, 10, and 12?

   Are there any two people in the class who have identical characteristics? Did you know that there are 64 possible combinations of these six characteristics?

4. You have looked at only six characteristics. There are hundreds of other characteristics you might have considered. What are some of these other characteristics?

Name ______________________ Date ____________ Class ____________

Chapter 6
Activity Worksheet

Classifying the Diversicules

Complete this activity after completing Lesson 2, which begins on page 132 of your textbook.

Meet the Diversicules family below. These little organisms are all mixed up and need you to help put them in order. Remember how you classified other living things. That's what you need to do with the Diversicules. Be patient! This is not an easy task. Here is a start:

(1, 2, 3, 4, 5, 6, 7, 8, 9, 10, 11, 12, 13, 14, 15)

(1, 3, 5, 7, 11, 12, 15) (2, 4, 6, 8, 9, 10, 13, 14)

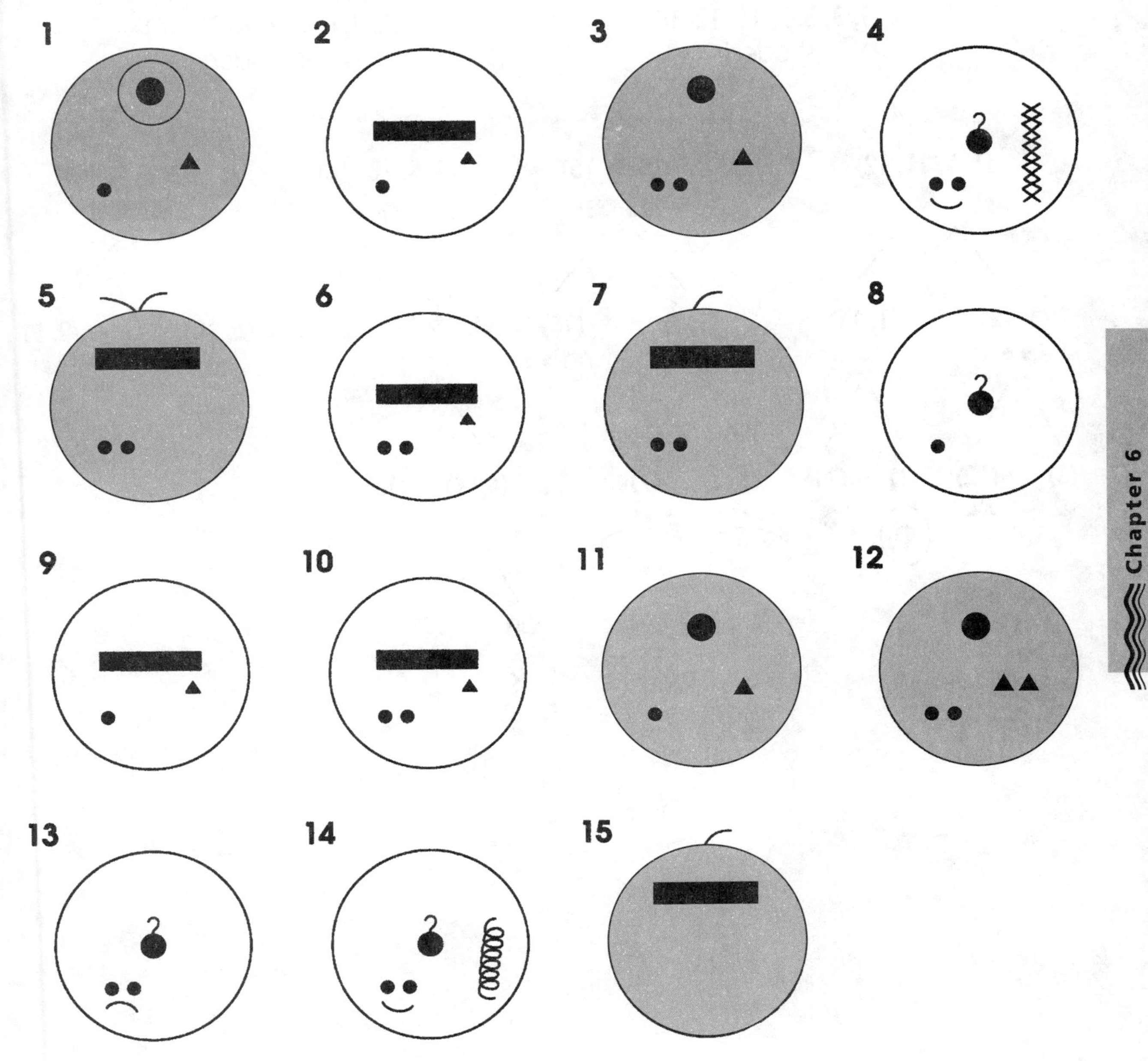

Name ______________________ Date ____________ Class ____________

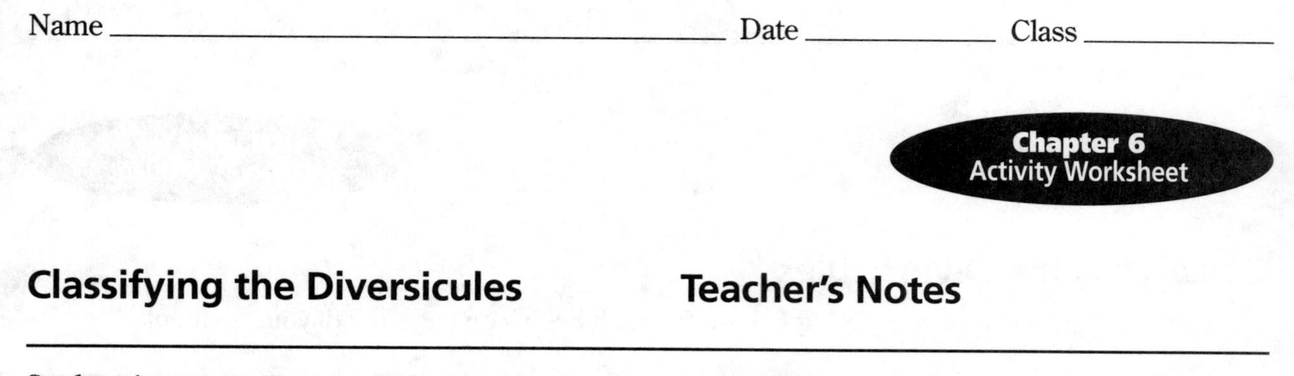

Classifying the Diversicules Teacher's Notes

Students' answers may vary. Here is one example of how the diversicules can be classified.

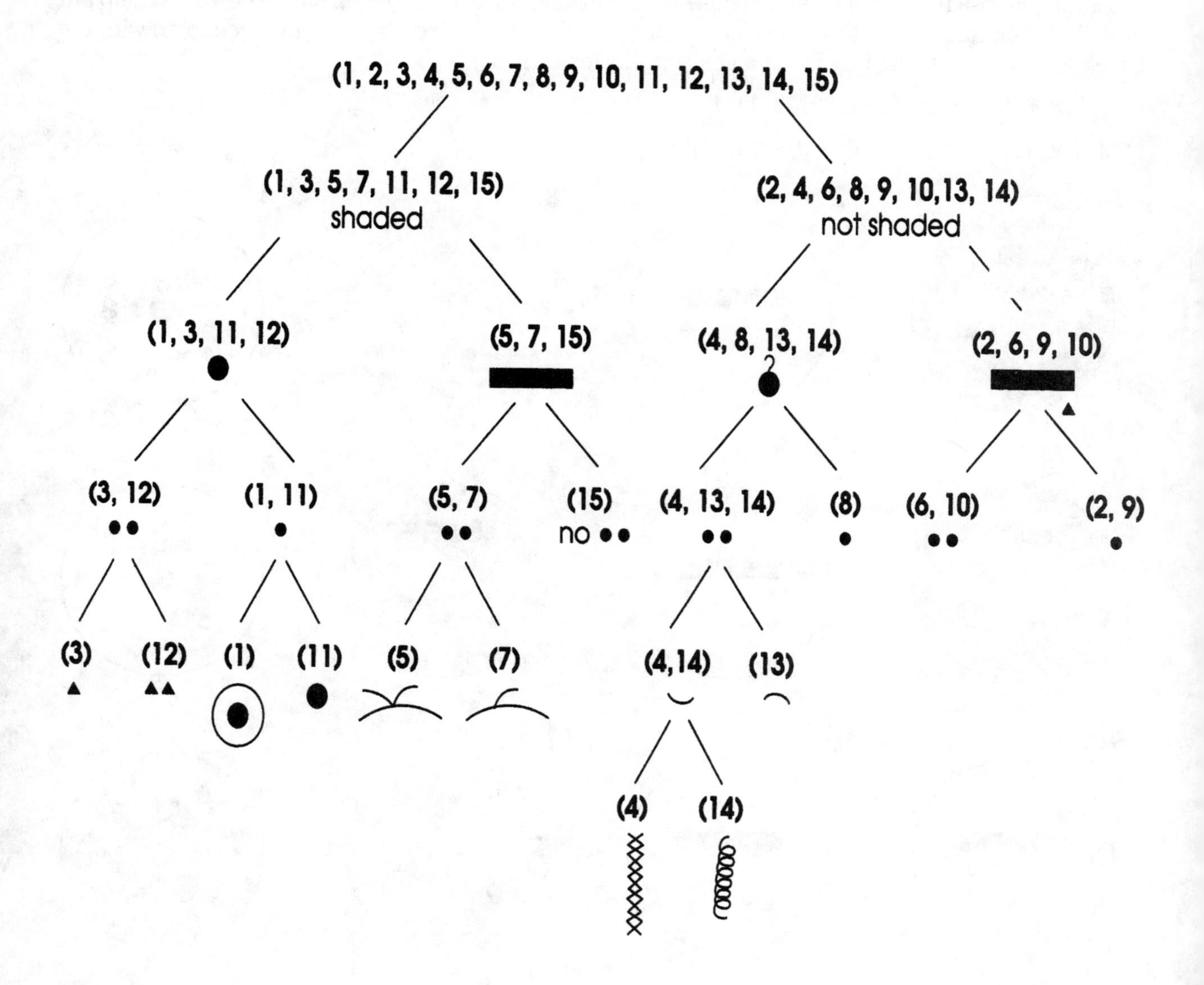

Name ______________________ Date ____________ Class ____________

Chapter 6
Review Worksheet

Challenge Your Thinking, page 136

1. The Animal-Kingdom Pie Can you explain the meaning of this pie chart? Which animals would you place in the missing piece of pie?

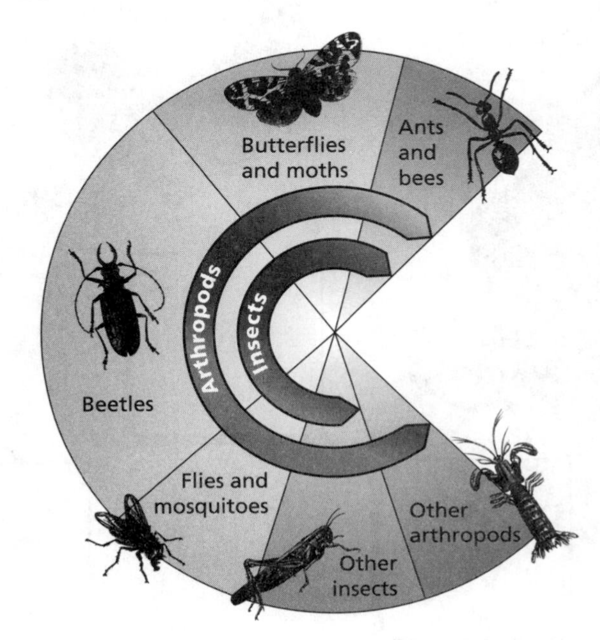

____

____

____

____

____

____

____

____

____

____

____

____

Chapter 6

Name ______________________ Date __________ Class __________

2. Birds of a Feather

Classify the birds shown according to their characteristics. Divide them into two groups, and write the characteristic shared by each group in the boxes of the first row. Then divide each of these groups into two subgroups, and write the characteristic shared by each subgroup in the boxes of the second row. Under each of the four boxes in the second row, write the names of the birds that fit into that subgroup.

Name ______________________ Date ____________ Class ____________

3. The Name Game

When scientists discover an organism, they may choose a name for a number of reasons. The name may reflect a characteristic of the species or the location where it was found. The name might even incorporate the name of a well-known scientist.

a. Shown here are photographs of organisms with their common name. Try to match each organism with its scientific name below. On what basis did you make each match?

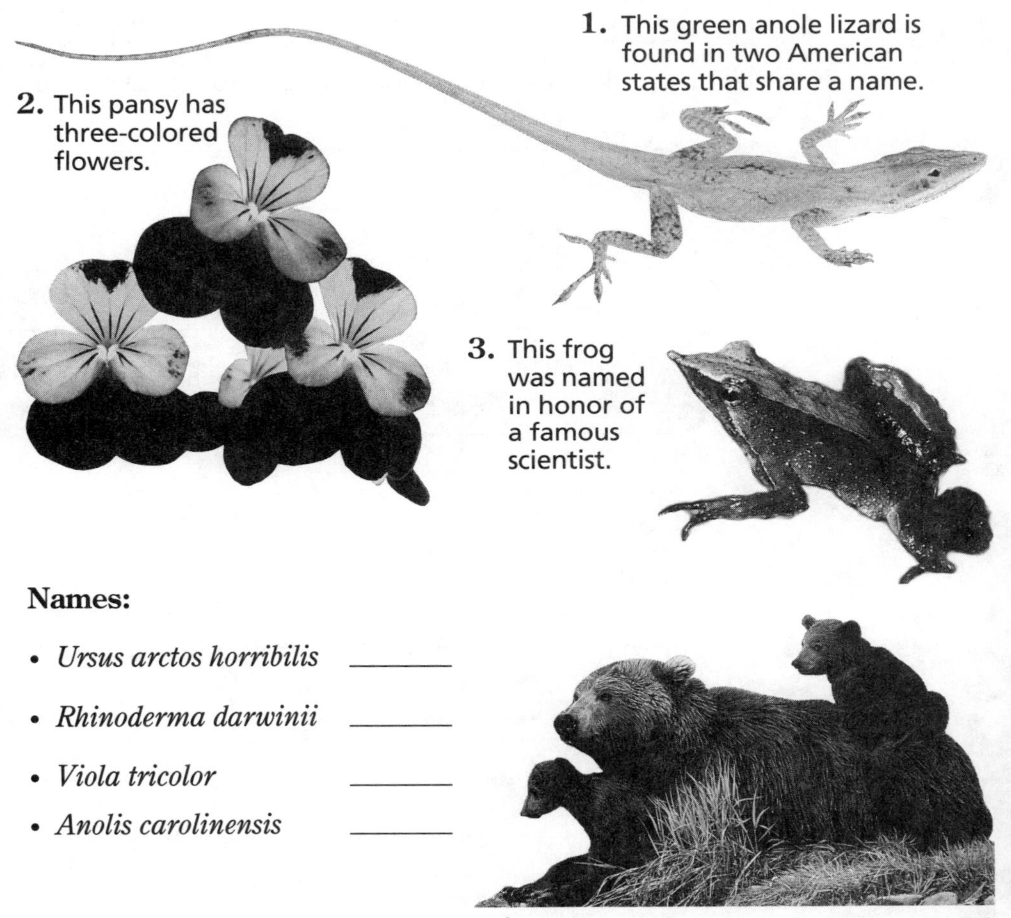
1. This green anole lizard is found in two American states that share a name.
2. This pansy has three-colored flowers.
3. This frog was named in honor of a famous scientist.
4. The grizzly bear fiercely protects itself, its family, and its food.

Names:

- *Ursus arctos horribilis* ______
- *Rhinoderma darwinii* ______
- *Viola tricolor* ______
- *Anolis carolinensis* ______

Photos also on page 137 of your textbook

b. Create a scientific name. Choose your favorite animal, and find out its scientific name. Now describe an imaginary animal that might be classified in the same genus. Complete the animal's scientific name by creating a species name. Give your reasoning for the name you chose.

____

____

____

____

Name ______________________ Date __________ Class __________

4. Can You Relate?

According to the classification systems used by biologists, why is

a. a turtle more closely related to an alligator than to an eel?

____

____

____

____

b. an earthworm more closely related to a grasshopper than to a snake?

____

____

____

____

c. an octopus more closely related to a clam than to a lobster?

____

____

____

____

d. a whale more closely related to a tiger than to a shark?

____

____

____

____

e. a maple tree more closely related to a rosebush than to a pine tree?

____

____

____

____

Name ______________________ Date ____________ Class ____________

Chapter 6
Assessment

Word Usage

1. Use all of the following terms in one or two sentences to show how they are related:

a. animal, subgroup, Linnaeus, vertebrates, kingdom, classification, group

____

____

____

____

b. arthropods, lobster, invertebrates, subgroup, kingdom

____

____

____

____

____

Correction/ Completion

2. The statements below are either incorrect or incomplete. Your challenge is to make them correct and complete.

a. *Homo sapiens* is the scientific name for ______________________.

b. According to modern biologists, mushrooms, mosses, and maples all belong to the plant kingdom.

____

____

____

Short Responses

3. For each animal below, write either *V* for vertebrate or *I* for invertebrate.

______ frog ______ shark ______ mosquito

______ lobster ______ bat ______ snail

Chapter 6

4. Circle the two organisms in each group that are most closely related to each other. Then tell what important characteristic they have in common.

   a. wasp earthworm eel ______________________

   b. seal tuna horse ______________________

   c. rosebush oak fern ______________________

   d. robin bat armadillo ______________________

CHALLENGE

Illustration for Interpretation

5. Look at the marine environment below and then use the space that follows to make a classification diagram for the organisms shown.

CHALLENGE 2

Short Essay

6. Why do scientists find it useful to have a consistent system for classifying living things?

Name ______________________ Date ____________ Class ____________

Test Your Memory

Use this activity as you conclude Unit 2.

Search through Unit 2 to locate each word for the given definition. The numbers under the blanks will help you complete the poem on the next page.

a. soft-bodied animals, most of them having shells: ___ ___ ___ ___ ___ ___ ___(21) ___

b. animals with feathers: ___ ___ ___ ___(1) ___

c. most mammals are born live, but reptiles, birds, fish, and amphibians develop from: ___ ___ ___(14) ___

d. simple land plants: ___ ___ ___ ___ ___(4) ___

e. a group of animals or plants that have certain permanent characteristics in common:

___ ___ ___ ___(19) ___ ___ ___

f. adaptations are what allow an organism to: ___ ___ ___(5) ___ ___ ___ ___

g. animals without backbones: ___ ___ ___ ___ ___ ___ ___ ___(16) ___ ___ ___ ___ ___

h. the Swedish scientist who devised a scientific way of classifying living things by their similarities:

___ ___ ___(10) ___ ___ ___ ___ ___

i. animals in this subgroup of arthropods have six legs and wings: ___(2) ___ ___ ___ ___ ___ ___

j. an animal with soft damp skin: ___ ___ ___(18) ___ ___ ___ ___ ___ ___

k. animals with backbones: ___(3) ___ ___ ___ ___ ___ ___ ___ ___ ___ ___

l. young grasshoppers whose diet determines their color: ___ ___(8) ___ ___ ___ ___

m. worms with many segments: ___ ___ ___(10) ___ ___ ___ ___ ___

n. the scientific name for human beings: ___ ___ ___ ___   ___ ___ ___(18) ___ ___ ___ ___

o. animals with scales and fins: ___(13) ___ ___ ___(15)

p. the person responsible for the theory of natural selection: ___ ___ ___ ___ ___ ___ ___(6)

___ ___ ___ ___ ___ ___

q. animals with jointed feet or legs: ___ ___ ___ ___ ___ ___(9) ___ ___ ___ ___

r. features of organisms that enable them to survive and reproduce:

___ ___ ___ ___ ___(7) ___ ___ ___ ___ ___ ___

Name ______________________ Date ____________ Class ____________

s. an imitative, attention-getting behavior: ___ ___ ___ ___ ___ ___ ___ (17)

t. Darwin's explanation of how the features of a species can change over many generations:

___ ___ ___ ___ ___ ___ ___ ___ ___ ___ ___ ___ ___ ___ ___ ___ ___ ___ (20, 11)

u. animals with fur or hair: ___ ___ ___ ___ ___ ___ ___ (12)

v. animals with scales but no fins: ___ ___ ___ ___ ___ ___ ___ ___ (7)

A Poem

Using the appropriate letters for the code numbers from the definitions above and on the previous page, fill in the blanks to discover a short poem.

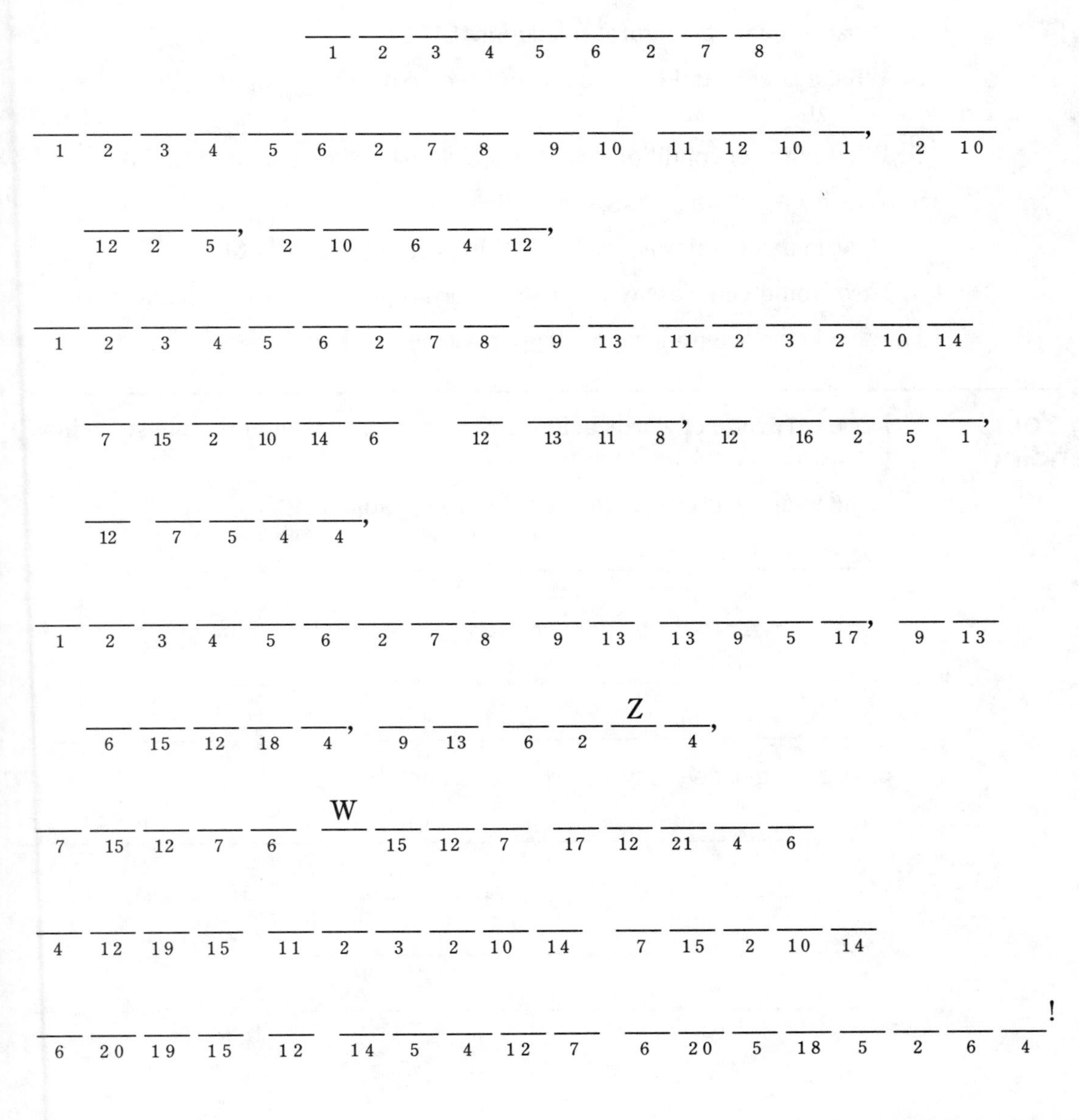

Name ______________________ Date ____________ Class ____________

Making Connections, page 138

The Big Ideas

In your ScienceLog, write a summary of this unit, using the following questions as a guide:

1. What is diversity? (Ch. 4)
2. Why is there diversity among living things? (Ch. 4)
3. What are some adaptations animals have for protection? for obtaining food? for attracting a mate? for locomotion? (Ch. 5)
4. How are plants adapted to survive and reproduce? (Ch. 5)
5. What is meant by natural selection? (Ch. 5)
6. What evidence did Darwin use to develop his theory of how life evolved? (Ch. 5)
7. What kinds of conditions cause species to become extinct? (Ch. 5)
8. Why do we classify things? (Ch. 6)
9. How, in a general way, are living things classified? (Ch. 6)
10. How would you classify yourself, according to Linnaeus's system? (Ch. 6)
11. What is the scientific name of your species? (Ch. 6)

Checking Your Understanding

1. For each group of words below, write one or two sentences that show how the words are related to each other.

   a. adaptation, predator, peppered moths, habitat

   ____

   ____

   ____

   ____

   b. organisms, diversity, environment, survival

   ____

   ____

   ____

   ____

Name ______________________ Date ____________ Class ____________

c. diversity, species, inherited characteristics, unique

____

____

____

____

2. Classify the living things in the photos on page 139 of your textbook according to some consistent system. Explain the system you used.

3. concept map Draw a concept map that shows how the following words are related to each other: coral snake; camouflage; mimicry; snowshoe hare; adaptations; snow; grasses, trees, and weeds; brown in summer; white in winter; and king snake.

Name ______________________ Date __________ Class __________

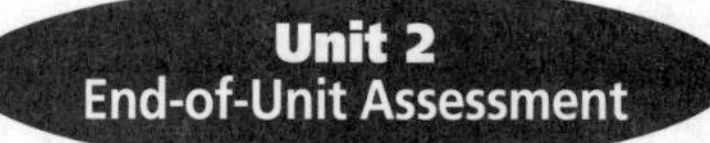

Word Usage

1. Use all of the following terms in one or two sentences to show how they are related.

   a. Charles Darwin, finches, natural selection, adaptations

   b. attract, hornet, mate, sting, peacock, adaptations, predators

2. The woolly mammoth was an elephantlike animal with a heavy coat. It lived in North America about 10,000 years ago, until the end of the last ice age. Use the terms below to make a hypothesis about the disappearance of these animals.

   extinct, environmental change, natural selection, adapt

Name ______________________ Date __________ Class __________

Correction/ Completion

3. The statements below are either incorrect or incomplete. Your challenge is to make them correct and complete.

   a. Arthropods are successful vertebrates because they survive in many different habitats.

   ____

   ____

   ____

   ____

   b. Scientists use a ______________ system to show the evolutionary relationship of organisms.

Short Responses

4. Using an example, describe an adaptation that a plant or animal has for the following:

   a. "disappearing" into its environment

   ____

   ____

   ____

   ____

   ____

   ____

   ____

   b. storing water

   ____

   ____

   ____

   ____

   ____

   ____

   ____

5. Seeds can be transported by wind, water, animals, or mechanical propulsion (like a bullet). Describe an adaptation that a seed might have for transportation by one of these methods.

____

____

____

____

____

____

____

6. How might Darwin have used the following observations to support his theory of natural selection?

   a. Finches that live in different parts of the Galápagos Islands have different kinds of beaks.

   ____

   ____

   ____

   ____

   ____

   ____

   ____

   b. Fossilized seashells are found high in the Andes Mountains of South America.

   ____

   ____

   ____

   ____

   ____

   ____

   ____

Data for Interpretation

7. In the table below are some characteristics for students in an eighth-grade class. Use the table to answer the questions that follow.

Characteristic		Van	Tasha	Tiffany	Bo	Josh
Hand folding, thumb position	left over right		X		X	X
	right over left	X		X		
Ear lobes	attached	X	X			X
	free			X	X	
Hairline	pointed		X		X	X
	straight	X		X		
Tongue rolling	can roll	X	X			
	can't roll			X	X	X

a. How many have three identical characteristics? Who are they?

____

____

b. How many have only two identical characteristics? Who are they?

____

____

c. How many have only one identical characteristic? Who are they?

____

____

d. Are there any two people who are different in every respect? If so, who are they?

____

____

Name ______________________________ Date ____________ Class ____________

e. How does this data relate to diversity?

____

____

____

____

____

Graph for Interpretation

8. The graph below estimates the number of named animal species in the world. Use the graph to answer the questions that follow.

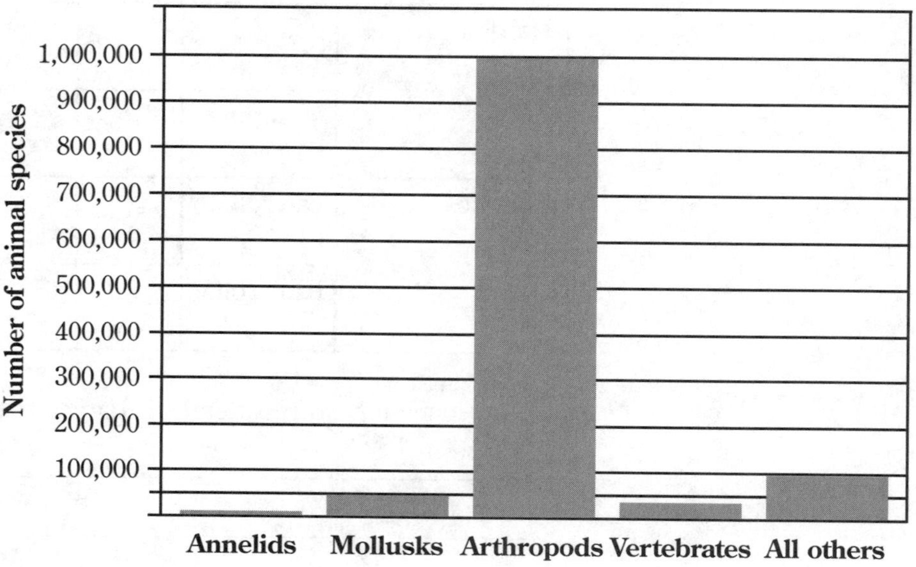

Number of animal species
1,000,000
900,000
800,000
700,000
600,000
500,000
400,000
300,000
200,000
100,000
Annelids
Mollusks
Arthropods
Vertebrates
All others

a. What is the total number of different kinds of animals?

____

____

b. Which animals are in the majority?

____

____

c. There are about ____________ times as many arthropods in the world as all other animals put together.

Name ______________________ Date __________ Class __________

d. What can you conclude about the diversity of the animals in this graph?

CHALLENGE

Numerical Problem

9. Suppose that the diversity of life on Earth were measured by percentages and that the 1.4 million known species correspond to a diversity level of 100 percent.

a. What does this mean?

b. There are 250,000 species of angiosperms, or flowering plants. What percentage of the known existing species on Earth do angiosperms make up?

c. How would the destruction of all angiosperms affect the diversity level of life on Earth?

Name ______________________ Date ____________ Class ____________

Illustration for Interpretation

10. Describe how each of these animals is adapted to survive in its environment. Mention adaptations for obtaining food, for protection, or for locomotion.

a. b. c.

a. ____

____

____

____

____

b. ____

____

____

____

____

c. ____

____

____

____

____

Name ______________________ Date __________ Class __________

CHALLENGE 2

Short Essay

11. What are some of the ways in which you benefit from the diversity of life?

____

____

____

____

____

____

____

____

____

Illustration for Interpretation

12. Examine the living things pictured below, and answer the questions that follow.

Sunflower

Butterfly

Cow

Rosebush

Earthworm

Pine tree with cones

Ostrich

Fern

a. On the next page, create a chart to classify these organisms into groups and subgroups. Then divide each of the subgroups so that each organism has its own subgroup. Explain the characteristic that distinguishes each organism in your classification system.

Your Classification Chart:

b. According to classification systems used by biologists today, why is a butterfly more closely related to an earthworm than to an ostrich?

Name ______________________ Date ____________ Class ____________

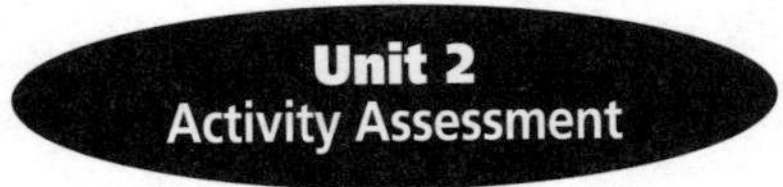

Exploring Diversity

Teacher's Notes

Overview	Students analyze different kinds of leaves and classify them by their similarities. They then create a poster detailing their classification system.	**Safety Alert!**

Materials
(per activity station)

- 8–10 different kinds of plant leaves (consider using 3 leaves from trees, 3 from shrubs or bushes, and 3 from fruits or vegetables)
- scissors
- colored pencils
- wax paper
- a metric ruler
- transparent tape
- a magnifying glass
- poster board

Preparation

Prior to the assessment, equip student activity stations with the materials needed for each experiment. Remind students not to taste the plant leaves.

Time Required

Each student should have 15 minutes at the activity station and 30 minutes to complete the Data Chart and poster.

Performance

At the end of the assessment, students should turn in the following:

- a completed Data Chart
- a poster detailing a classification system

Evaluation

The following is a recommended breakdown for evaluation of this Activity Assessment:

- 10% use of equipment and preparation of leaves
- 40% quality and clarity of observations
- 40% design of a clear classification system
- 10% neatness and clarity of poster

Name ______________________ Date ____________ Class ____________

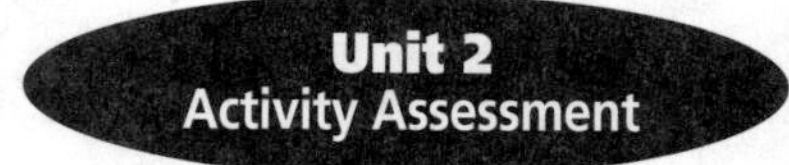

Exploring Diversity

Imagine that you have been transported back in time to the eighteenth century. You have been selected to work with scientist Carolus Linnaeus as he develops the first classification system for plants and animals. You've been asked to examine several plant leaves and to decide which factors are important in classifying them. What recommendations will you make to Linnaeus?

Before You Begin . . .

As you work through the tasks, keep in mind that your teacher will be observing:

- how you prepare the leaves for examination
- your ability to record your observations clearly
- how well you classify the leaves by their similarities
- how clearly you present your classification system on the poster

Now you are ready to begin your eighteenth-century adventure!

Task 1

Cut the wax paper into sheets. Press one leaf between two sheets of wax paper and tape the edges of the papers together. Repeat this procedure for all of the leaves at your workstation. Then assign each leaf a number so that you can distinguish each of your samples.

Task 2

Examine each leaf. What are its basic characteristics? Record your observations on your Data Chart.

Task 3

Analyze your Data Chart for similarities among the leaves. Classify the samples by placing similar leaves in the same group.

Task 4

Outline your classification system on a piece of poster board. Tape similar samples next to each other with a brief explanation of what the leaves have in common and how they are different from the rest of the leaves.

Name ______________________ Date ____________ Class ____________

Activity Assessment, continued

Classification System Data Chart

Leaf number	Color	Size	Shape	Other features

Name ______________________ Date __________ Class __________

Unit 2
Self-Evaluation

Self-Evaluation of Achievement

The statements below include some of the things that may be learned when studying this unit. If I have put a check mark beside a statement, that means I can do what it says.

_____ I can describe differences in shape, size, and structure of living things. (Ch. 4)

_____ I can explain the need for diversity among living things. (Ch. 4)

_____ I can give examples of how plants and animals are adapted to their environment. (Ch. 5)

_____ I can explain how adaptations help living things survive in their environment. (Ch. 5)

_____ Using examples, I can explain the theory of natural selection. (Ch. 5)

_____ I can explain the effects, both negative and positive, that humans have had on the extinction of organisms. (Ch. 5)

_____ Using a chart, I can classify both plants and animals into two major groups each and give examples of each group. (Ch. 6)

_____ I can explain why biologists use systems to classify living things. (Ch. 6)

I have also learned to ______________________

____

____

____

____

____

I would like to know more about ______________________

____

____

____

____

____

Signature: ______________________

Name ______________________ Date __________ Class __________

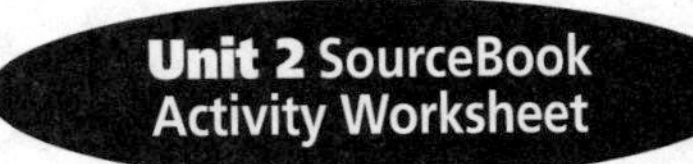

SourceBook

Mountain Mice

Complete this activity after reading pages S41–S43 of the SourceBook.

Background

You are a member of a zoological team exploring ecosystems in the mountains. Scientists believe that, over a long period of time, a particular mouse species migrated to the mountains from a valley over 20 km away. Your job is to evaluate the mice that are found in the mountains at various altitudes.

- At the base of the mountain, you find plenty of mice. They have brown or gray coats, short fur, and large ears. They nest in hollow logs, beds of leaves, deep grass, and almost any other place that offers shelter. In your population samples, you find that there is an average of 200 mice per square kilometer. You observe mice eating nuts, grasses, leaves, insects, berries, and litter from garbage cans placed along the roadways.
- Halfway up the mountain, at about 3000 m, you find an average of 120 mice per square kilometer. Their coats vary from light tan to mottled brown, they have medium-sized ears, and the length of their fur varies from population to population. Most of the mice nest at the base of fir trees among broken branches, cones, and fallen needles. Their primary food sources are grasses, nuts, bark, and insects.
- At 5500 m, the mouse population drops significantly. You find an average of about 40 mice per square kilometer. These mice have small bodies and small ears. They have long, white coats. Their nests lie beneath outcroppings of rocks or deep under the snow-covered ground, and they eat bark and insects buried beneath the ground.

What to Do

Use the chart on the next page to show how the mice have adapted to the different mountain environments. Complete the chart as follows:

Column 1: List each kind of mouse and briefly identify where it makes its home on the mountain.

Column 2: Describe at least two adaptations that help each kind of mouse survive in its environment.

Column 3: Tell what environmental condition each adaptation was in response to.

Finally, draw some conclusions about your analyses. What characteristics vary among the mouse populations? When mice first migrated to the mountain range, what do you think they looked like? How could natural selection and adaptation have affected that original population?

Name ________________ Date ________ Class ________

SourceBook Activity Worksheet, continued

Mouse and habitat	Adaptations	Environmental conditions
Brown or gray mice at the base of the mountain	• Eat a wide variety of foods, including nuts, grasses, leaves, insects, berries, and food left behind by people • Short hair	• Abundant and varied food sources • Warm temperatures

Conclusions

____

____

____

____

____

____

____

____

____

Name ______________________ Date __________ Class __________

Unit 2 SourceBook Review Worksheet

Unit CheckUp, page S49

Concept Mapping

The concept map below illustrates major ideas in this unit. Complete the map by supplying the missing terms. Then extend your map by answering the additional question below.

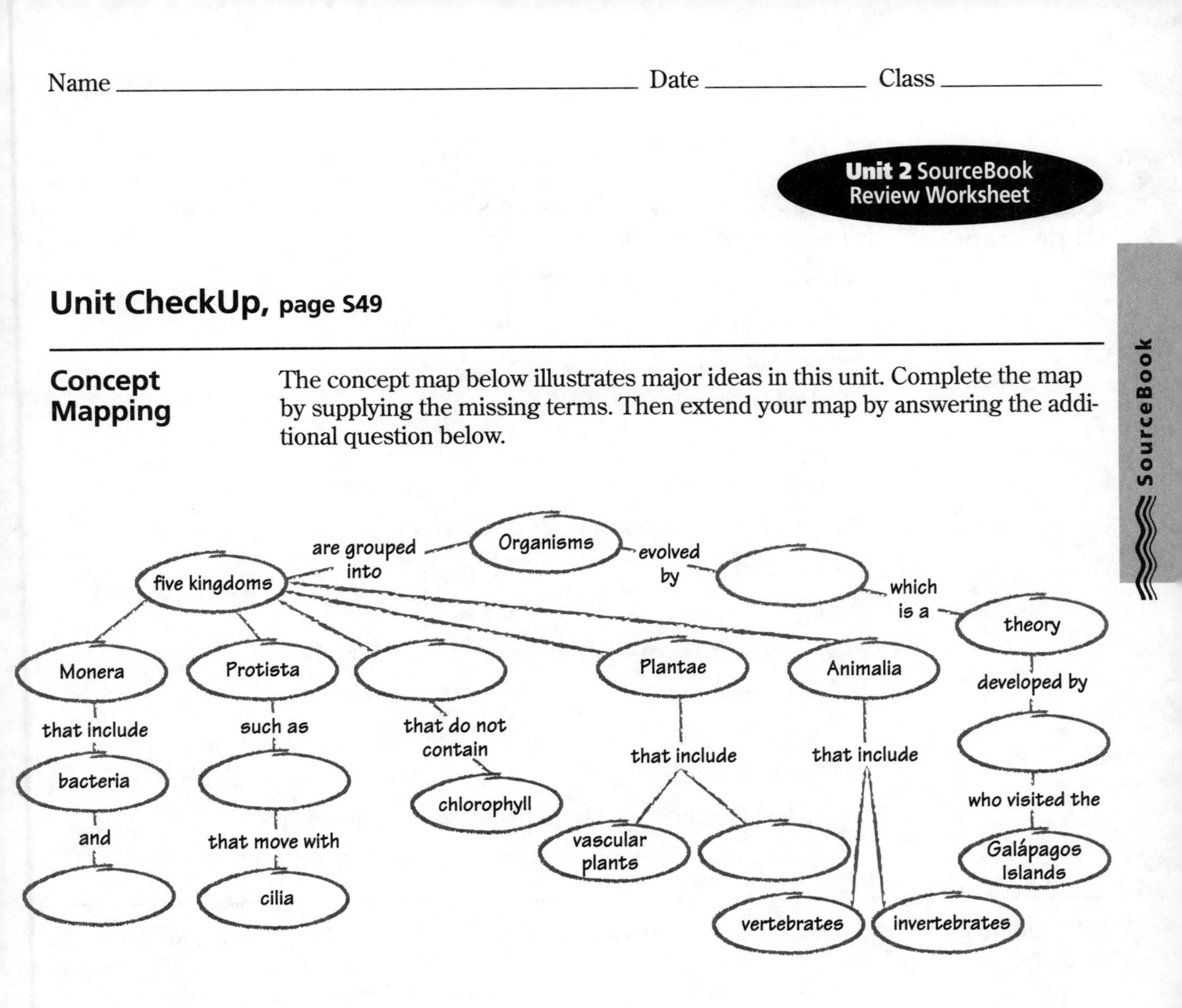

Where and how would you connect the terms *spores, algae,* and *backbones?*

Name ______________________ Date ____________ Class ____________

Checking Your Understanding

Select the choice that most completely and correctly answers each of the following questions.

1. Which is NOT a kingdom name?
   - **a.** Protista
   - **b.** Animalia
   - **c.** Protozoa
   - **d.** Fungi
2. An animal with a backbone is called
   - **a.** a segmented worm.
   - **b.** a protist.
   - **c.** an echinoderm.
   - **d.** a vertebrate.
3. Which is a component of fungi but NOT of plants?
   - **a.** chlorophyll
   - **b.** hyphae
   - **c.** a nucleus
   - **d.** a cell wall
4. During what interval did most of the Earth's history occur?
   - **a.** Cambrian era
   - **b.** Precambrian time
   - **c.** Mesozoic era
   - **d.** Paleozoic era
5. Similarly constructed limbs on different vertebrates are called
   - **a.** homologous structures.
   - **b.** biochemical similarities.
   - **c.** wings.
   - **d.** mutational structures.

Interpreting Graphs

Bacteria are grown in a test tube that contains a limited amount of food. This graph illustrates how the population of bacteria changes over a two-day period. What do you predict will happen to the size of the population after day two? Support your prediction with an explanation on the next page.

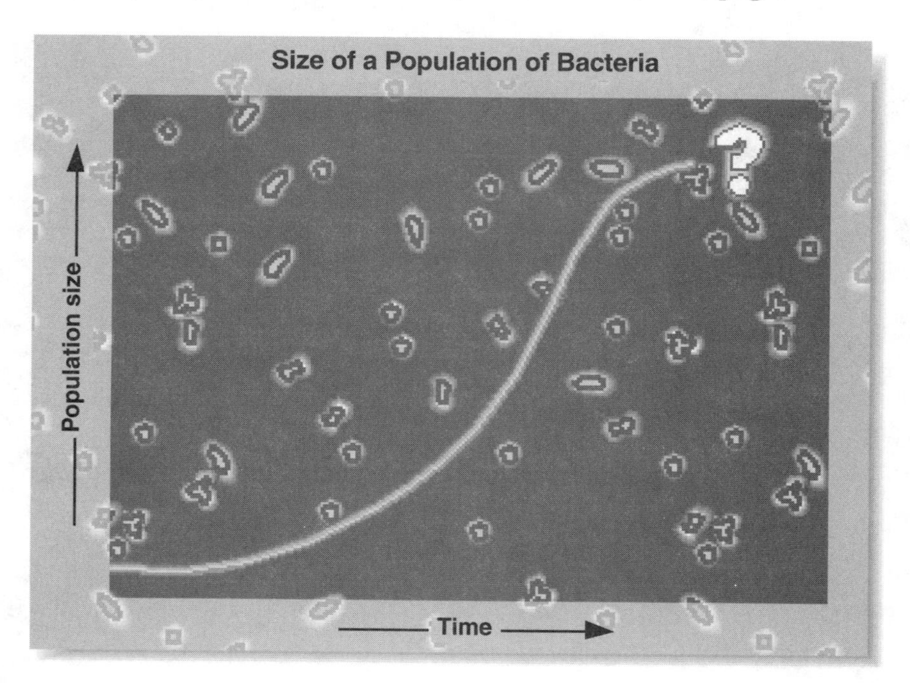

Graph also on page S50 of your textbook

SourceBook

Critical Thinking

Carefully consider the following questions, and write a response in the space below that indicates your understanding of science.

1. A biologist notices a large number of short plants growing near a swampy area. Although she thinks they are nonvascular plants, she is not sure. She collects one of the plants to bring back to her laboratory, where she will look at a slice from its stem under a microscope. Why does she wish to observe the plant's stem? Why is her guess that the plants are non-vascular a logical assumption?

2. Explain why an earthworm is able to crawl faster than a planarian can move.

3. There are many species of small birds called finches that live on the Galápagos Islands. Scientists think that the many different kinds evolved from a single species of finch that flew to the islands years ago. Is it more likely that this ancestral species arrived from Europe, North America, or South America? Explain the reason for your choice.

4. You discover a new species of protozoan but are unsure where to place it on the protist evolutionary tree. What kind of observations would you make about the protist, and what kind of analyses would you perform on it to help you classify this organism?

5. The wing of a bird and the wing of an insect have the same function, yet they are NOT homologous structures. Explain why they are not.

Portfolio Idea In your ScienceLog, create a table that summarizes the major characteristics of the organisms in each of the five kingdoms. Start by placing the kingdom names as headings in your table. Then compare and contrast the organisms in each kingdom to come up with the characteristics you will have in your table. Make your table large enough to include a colored drawing or a magazine clipping of a representative organism for each kingdom.

Name ______________________ Date __________ Class __________

Unit 2 SourceBook Assessment

1. Fill in the blanks with the following subcategories, in order, beginning with kingdom and ending with the most specific classification: class, family, species, phylum.

   kingdom, ______________, ______________, order,

   ______________, genus, ______________

2. Put the following terms in the proper order, from the longest period of time to the shortest: era, decade, period, week, year.

   ____

3. Monerans, protists, animals, and fungi are examples of the category

   a. phylum. **b.** class. **c.** species. **d.** kingdom.

4. Cyanobacteria contain chlorophyll and produce their own food by a process called

   a. respiration. **b.** photosynthesis. **c.** conduction. **d.** osmosis.

5. Protozoans are animal-like protists; therefore, they (do/do not) have a cell wall, and they (do/do not) contain chlorophyll. (Circle the correct words.)

6. The black or gray powdery substance that you may find growing on an old hamburger bun is an example of a(n)

   a. fungus. **b.** cyanobacterium. **c.** alga. **d.** moneran.

7. Redwood trees are

   a. vascular plants. **b.** nonvascular plants.

8. The most diverse group of modern plants are those with

   a. "naked" seeds. **b.** "covered" seeds.

9. Which of the following does NOT correctly complete the sentence? You are

   a. an invertebrate. **b.** an animal. **c.** warmblooded. **d.** a mammal.

10. Animals must

    a. be multicellular. **b.** have organ systems. **c.** have vertebrae.

    d. be able to make their own food within their body.

    e. All of the above **f.** **a** and **b** only **g.** **a**, **b**, and **d** only

11. A man walked into a restaurant and ordered an arthropod. Which of the following did he order?

    a. steak **b.** catfish **c.** lobster **d.** chicken

12. A person's bones are called an

a. endoskeleton. **b.** exoskeleton.

13. Coldblooded refers to

a. animals having a body temperature that changes according to the temperature of the environment.

b. all vertebrates that live in the water.

c. animals having a body temperature below 20°C.

d. **a** and **b** only

14. Which invertebrates have gills when they are young and lungs when they mature? These animals also live part of their lives in water and part on land.

a. reptiles **b.** coelenterates

c. amphibians **d.** segmented worms

15. The term ***millennium*** means

a. 10,000 years. **b.** 5 decades.

c. any long period of time. **d.** 1000 years.

16. A blue jay laid two eggs. The first egg hatched a very strong and healthy female blue jay. The second egg hatched a much weaker and smaller female blue jay. The first blue jay was able to learn to fly, mate, and reproduce. The second was unable to fly very well and died in just a few weeks. What process is described here?

a. fate **b.** natural selection **c.** adaptation **d.** mutation

17. Farmer Jane grew tomatoes. She took the seeds from the largest, reddest, and best-tasting tomatoes and planted them. From the plants that grew, she again took the seeds from the largest, reddest, and best-tasting tomatoes and planted them. She repeated this for three seasons. Farmer Jane then had the largest, reddest, and best-tasting tomatoes in the whole county. This scientific process is known as

a. survival of the fittest. **b.** resistance.

c. selective breeding. **d.** variation.

18. Fishermen once tried to eliminate starfish by cutting them into pieces and throwing them back into the sea. This, however, did not solve their problem. It only made it worse. What characteristic of starfish (an echinoderm) explains why the situation worsened?

____

____

____

19. If you were digging deep in your backyard and found a piece of stone with the imprint of a starfish, could you find a scientific explanation for how it got there? (Your neighbor burying it there is not a scientific explanation.)

20. Why are albino Bengal tigers more scarce than non-albino Bengal tigers? (Assume that albinism is just as common as non-albinism in Bengal tigers.)

21. The bones of a person's arm and the bones of a whale's flipper are arranged similarly, but they differ in size and shape. What can you infer from this observation?

22. Only mammals are warmblooded.

a. true **b.** false

23. Birds have hollow bones that better enable them to fly.

a. true **b.** false

24. From what we have learned about the history of the Earth, mammals were among the first organisms to appear on Earth.

a. true **b.** false

25. All organisms in the kingdom Protista are multicellular.

a. true **b.** false

Estimado padre/madre de familia,

En las próximas semanas, su hijo(a) explorará la diversidad de los seres vivos y las razones para la misma. Estudiará también el modo cómo se han adaptado a sus hábitats y nichos particulares las diferentes plantas y los animales. Se examinará la clasificación de organismos vivos. Cuando los estudiantes hayan terminado la Unidad 2, deberán ser capaces de dar respuesta a las siguientes preguntas, para captar las "grandes ideas" de la unidad.

1. ¿Qué es la diversidad? (Cap. 4)
2. ¿Por qué hay diversidad entre los seres vivos? (Cap. 4)
3. ¿Puedes decir cuáles son algunas de las adaptaciones por las que pasan los animales para protegerse?, ¿para conseguir alimento?, ¿para atraer a su pareja?, ¿para la locomoción? (Cap. 5)
4. ¿Cómo se adaptan las plantas para sobrevivir y reproducirse? (Cap. 5)
5. ¿Qué es la selección natural? (Cap. 5)
6. ¿Qué pruebas usó Darwin para desarrollar su teoría de cómo evolucionó la vida? (Cap. 5)
7. ¿Qué condiciones hacen que se extingan las especies? (Cap. 5)
8. ¿Por qué clasificamos las cosas? (Cap. 6)
9. ¿Cómo se clasifican los seres vivos en términos generales? (Cap. 6)
10. ¿Cómo te clasificarías a ti mismo según el sistema de Lineo? (Cap. 6)
11. ¿Cuál es el nombre científico de tu especie? (Cap. 6)

Si Ud. quiere, puede hacerle a su hijo(a) algunas de estas preguntas a medida que vamos avanzando en la unidad. Esto será para él/ella un valioso modo de practicar la comunicación de sus conocimientos. No se preocupe si las respuestas no se oyen como si se hubieran memorizado en un libro de texto. Con este libro, los estudiantes van a descubrir las respuestas por sí mismos, individualmente o en grupos. Esto es al mismo tiempo el reto que les presenta la ciencia y el placer que les da.

Atentamente,

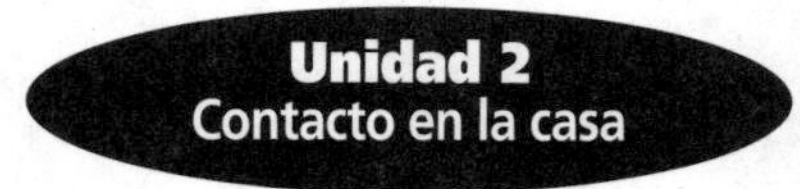

Los materiales que aparecen abajo van a ser usados en clase para varias exploraciones de ciencia de la Unidad 2. Su contribución de materiales va a ser muy apreciada. He marcado algunos de los materiales en la lista. Si usted los tiene y quiere donarlos, por favor mándelos a la escuela con su hijo o hija para el

______________________________________.

- ○ cinta adhesiva
- ○ cartón o papel grueso
- ○ palitos para artesanías (craft sticks)
- ○ dados
- ○ guías de viaje para identificar plantas
- ○ plantas verdes
- ○ tarjetas índice
- ○ leños (chicos)
- ○ lupas
- ○ marcadores
- ○ cinta para medir (metrica)
- ○ envoltura de plástico
- ○ grava blanda (pulverizada, como gis)
- ○ tierra
- ○ cucharas
- ○ palitos
- ○ cordón (grueso)
- ○ lana

Desde ya, le agradecemos su ayuda.

En la Unidad 2, Diversidad de los seres vivos, vas a investigar la diversidad entre los seres vivos y la razón de la misma. Examinarás cómo se han adaptado a sus hábitats y nichos particulares las diferentes plantas y animales. Vas a leer sobre las polillas moteadas de Inglaterra y sobre los pinzones de las islas Galápagos. Al ir leyendo la unidad, trata de responder a las siguientes preguntas. Estas son las "grandes ideas" de la unidad. Cuando puedas contestar estas preguntas, habrás logrado entender bien los principales conceptos de esta unidad.

1. ¿Qué es la diversidad? (Cap. 4)
2. ¿Por qué hay diversidad entre los seres vivos? (Cap. 4)
3. ¿Puedes decir cuáles son algunas de las adaptaciones por las que pasan los animales para protegerse?, ¿para conseguir alimento?, ¿para atraer a su pareja?, ¿para la locomoción? (Cap. 5)
4. ¿Cómo se adaptan las plantas para sobrevivir y reproducirse? (Cap. 5)
5. ¿Qué es la selección natural? (Cap. 5)
6. ¿Qué pruebas usó Darwin para desarrollar su teoría de cómo evolucionó la vida? (Cap. 5)
7. ¿Qué condiciones hacen que se extingan las especies? (Cap. 5)
8. ¿Por qué clasificamos las cosas? (Cap. 6)
9. ¿Cómo se clasifican los seres vivos en términos generales? (Cap. 6)
10. ¿Cómo te clasificarías a ti mismo según el sistema de Lineo? (Cap. 6)
11. ¿Cuál es el nombre científico de tu especie? (Cap. 6)

Vocabulario

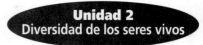

Adaptation (97)	**Adaptación** característica heredada que aparece con el tiempo y permite que los organismos sobrevivan mejor en un medio ambiente dado
Appendage (112)	**Apéndice** parte de un organismo que está añadido al cuerpo, como una cola o un dedo
Camouflage (98)	**Camuflaje** el método por el que un organismo se disfraza o se confunde con sus alrededores para esconderse de los predadores
Classify (115)	**Clasificar** separar en grupos
Diversity (78)	**Diversidad** diferencias o variedad entre los seres vivos
Endangered species (110)	**Especie en peligro de extinción** una especie en la cual el número de representantes ha bajado tanto que es posible que se extinga en un futuro cercano
Extinction (108)	**Extinción** la desaparición irreversible de una especie
Gene therapy (141)	**Terapia genética** el proceso de remplazar genes defectuosos en células, inyectando genes sanos a las células
Genetic disorder (141)	**Desorden genético** enfermedad causada por genes defectuosos en los cromosomas de un organismo
Invertebrate (120)	**Invertebrado** animal sin columna vertebral
Mimicry (101)	**Mimetismo** la semejanza en comportamiento, apariencia, sonido u olor de un organismo y otro organismo u objeto en sus alrededores. El mimetismo ayuda a proteger los organismos de los predadores.
Natural selection (92)	**Selección natural** la selección o el favorecimiento de los organismos que tienen la mayor capacidad de sobrevivir y reproducirse en su medio ambiente. Charles Darwin usó esta idea para explicar cómo las características de una especie pueden cambiar en muchas generaciones.
Organism (78)	**Organismo** ser vivo con órganos y partes que funcionan en conjunto
Scientific name (133)	**Nombre científico** el nombre único, de dos palabras, (correspondientes a los nombres de género y de especie del organismo) que los científicos dan a cada clase de organismo de la Tierra
Vertebrate (120)	**Vertebrado** animal con una columna vertebral

Photo/Art Credits

Abbreviated as follows: (t) top; (b) bottom; (l) left; (r) right; (c) center.

Photo Credits
Page i(l), Page Overtures; i(b), Jeff Smith/FotoSmith, special thanks to Reptile Solutions of Tucson; 3(bl), Phil A. Dotson; 5(tl), Susan Middleton/Tony Stone Images; 9(t-b), John Warden/Tony Stone Images; Mueller/ZEFA/H. Armstrong Roberts; Chris Collins/The Stock Market; Breck P. Kent; T. Ulrich/H. Armstrong Roberts; 10(t-b), Richard Kolar/Animals Animals; Suzanne L. Collins & Joseph E. Collins/Photo Researchers, Inc.; R. Kord/H. Armstrong Roberts; HRW photo by Daniel Schaefer; Rod Planck/Photo Researchers, Inc.; 31(both), Mickey Gibson/Animals Animals; 50(clockwise from top), Robert Maier/Animals Animals; Ralph A. Reinhold/Animals Animals; Jim Zipp//Photo Researchers, Inc.; Rod Planck/Photo Researchers, Inc.; L. West/Photo Researchers, Inc.; John Warden/Tony Stone Images; Doug Armand/Tony Stone Images; Bruce Coleman, Inc.; 51(l), Henley & Savage/The Stock Market; 51(tr), M.H. Sharp/Photo Researchers, Inc.; 51(cr), F. Gohier/Photo Researchers, Inc.; 51(br), Tony Dawson/Tony Stone Images; 87(l-r), John Warden/Tony Stone Images; Mueller/ZEFA/H. Armstrong Roberts; Chris Collins/The Stock Market; Breck P. Kent; T.Ulrich/H. Armstrong Roberts; 88(l-r), Richard Kolar/Animals Animals; Suzanne L. Collins & Joseph E. Collins/Photo Researchers, Inc.; R. Kord/H. Armstrong Roberts; HRW photo by Daniel Schaefer; Rod Planck/Photo Researchers, Inc.; 95(both), Mickey Gibson/Animals Animals; 102(tl), M.H. Sharp//Photo Researchers, Inc.; 102(tc), F. Gohier/Photo Researchers, Inc.; 102(tr), Tony Dawson/Tony Stone Images; 102(clockwise from l), Robert Maier/Animals Animals; Ralph A. Reinhold/Animals Animals; Jim Zipp/Photo Researchers, Inc.; Rod Planck/Photo Researchers, Inc.; L. West/Photo Researchers, Inc.; John Warden/Tony Stone Images; Doug Armand/Tony Stone Images; Bruce Coleman, Inc.

Art Credits
All work, unless otherwise noted, is contributed by Holt, Rinehart and Winston.
Page 6, Sarah Woodward/Morgan Cain & Associates; 13, Rhoda Grossman; 15, Debbie Peel; 19, Peter Van Gulik; 27, Paul Blakey/Susan Wells & Associates, Inc.; 29, Paul Blakey/Susan Wells & Associates, Inc.; 35, Donna Kae Nelson/Sharon Langley Artist Representatives; 36, Amy Wasserman; 47, Morgan Cain & Associates; 48, Morgan Cain & Associates; 49, Bruce Metherd; 54, Holly Cooper; 66, Russell Moore; 67, Morgan Cain & Associates; 76, Glasgow & Associates; 89, Rhoda Grossman; 90, Debbie Peel; 97(l), Donna Kae Nelson/Sharon Langley Artist Representatives; 97(r), Amy Wasserman; 101, Bruce Metherd; 104(lc), Holly Cooper; 110(l), Russell Moore; 110(r) Morgan Cain & Associates; 112(r), Glasgow & Associates

Cover Credits
DragonFly-Martin Wendler/Okapia/Photo Researchers, Inc.; Chameleon-Jeff Smith/FotoSmith/with special thanks to Reptile Solutions of Tucson; Assorted Textures-Page Overtures.

Answer Keys

Unit 2: Diversity of Living Things

Contents

Name ______ Date ______ Class ______

EXPLORATION 2

Chapter 4 Exploration Worksheet

The Puzzling Diversity of Living Things, page 82

Your goal to learn to identify organisms by looking closely at their specific characteristics

Activity 1: Who Am I?

In the following riddles, you'll find information about the structures, habits, and habitats of some organisms. Read each riddle carefully, and think about what it tells you. Which living thing does each riddle describe? If you're having trouble figuring out what a riddle is describing, look at the pictures on this page and the next for some hints. (The pictures are not in order.) If you are really stuck, the answers are in code following the riddles. All you have to do is break the code!

Brown bear

Rainbow trout

Onion

Apple

Woodpecker

Photos also on page 82 of your textbook

Riddle 1 Dragonfly

I move slowly when I am young but very quickly when I'm an adult. I eat flying insects, which I hunt near water. I have to be a strong flier to catch my food. When I stretch my four wings, I look like a helicopter. I have two more legs than a dog, and I have very large eyes. I am coldblooded and have an external (outside) skeleton. Sometimes I'm very colorful. Who am I?

Riddle 2 Human

I can walk, run, and swim. I can see well, but my sense of smell is not very sharp. I am warmblooded. I am very adaptable and can live in many different environments. I really enjoy changing my environment. I care for my young for many years. I stand upright. Who am I?

Riddle 3 Toad

I must live in damp or wet places, avoiding the dry heat of summer and the cold of winter. If I am living in a cold climate, I become dormant in the winter. If I'm a female, I produce young by laying eggs in water. I survive by eating any moving thing that I can swallow. I can sing very well. Some of my close relatives can secrete a sticky, white poison that can kill or paralyze dogs or other enemies who may try to eat them. Who am I?

Riddle 4 White pelican

I live in lakes, marshes, salt bays, and on beaches. I eat mostly fish and crustaceans. Although I can fly, I catch fish only by swimming. My great throat pouch is handy for scooping up fish. I fly by alternating several flaps of my wings with a glide. I always fly with my head hunched behind my shoulders. I nest on the ground in colonies. I have a wingspan of 2.5–3.0 m. My close cousins live only by the ocean, but I can venture inland. I am happy to report that these cousins are growing in number, even though they suffered from DDT poisoning a few years ago. Who am I?

Name ______________ Date ________ Class ________

Exploration 2 Worksheet, continued

White pelican

Toad

Elk

Human

Dragonfly

Photos also on page 83 of your textbook

Riddle 5 **Woodpecker**

I have a very high body temperature. My feet are well adapted for grasping things. I have four toes on each foot: two point forward, and two point backward. I have stiff, spiny tail feathers that act as a prop when I hunt food. I eat tree-boring insects, ants, acorns, flying insects, berries, and sap. My home, which I make myself, is a hole in a tree. I use my bill to chisel away the wood. Who am I?

Riddle 6 **Onion**

I have pointed, green stalks above the ground and a rounded, brown bulb below. People must pull me out of the soil before I can be useful to them. Cooks use me to improve the taste of food. If people bite me, I can bite back, making their eyes water. Who am I?

Riddle 7 **Rainbow trout**

I live in cold, well-oxygenated water, and I'm a fast, strong swimmer. I am slim, sleek, and colorful. I'm a carnivore; I eat mostly insects and smaller members of my own kind. I spawn my eggs during the spring in small, clear streams. I'm coldblooded. Who am I?

Riddle 8 **Apple**

I undergo wondrous changes during my life. At the beginning, I am a sweet-smelling, pink-and-white blossom. Later I'm a hard, green ball that makes your eyes water and your mouth pucker if you try to eat me. Finally, I become a sweet, juicy, red or yellow fruit. People say I keep physicians away. Who am I?

Riddle 9 **Brown bear**

I am a big animal. My mass is about 225 kg, but my tail is only about 15 cm long. I am dark in color. Generally, I live on forest floors and in thickets. When it starts to get really cold, I enter my shelter for the winter. I don't have very good sight, but my senses of hearing and smell are keen. Using these senses, I find lots of food—small animals, insects, garbage, leaves, grasses, berries, nuts, and fruits. Who am I?

Riddle 10 **Elk**

I am warmblooded and hairy. I feed milk to my young. I chew my cud, and I have a complex stomach. The males of my kind have huge, branching antlers. I have a heavily maned neck. Humans, wolves, and mountain lions are my only enemies, but mountain lions usually won't attack me when I'm fully grown. My young are not camouflaged from these enemies until their winter hair grows out. Sometimes you can hear the males of my kind give a high-pitched bugle call. If this call is answered by another male, a battle may follow. Who am I?

Name ______________ Date ________ Class ________

Exploration 2 Worksheet, continued

Here are the coded answers to the riddles:

1. CQZFNMEKX
2. GTLZM
3. SNZC
4. VGHSD ODKHBZM
5. VNNCODBJDQ
6. NMHNM
7. QZHMANV SQNTS
8. ZOOKD
9. AQNVM ADZQ
10. DKJ

Hint for decoding: ZMS = ANT

Activity 2: You Be the Riddler!

Read the riddles again carefully, this time noticing which characteristics of the organisms are used to describe their diversity. Then try writing your own riddles for some of the living things pictured on pages 84–85 of your textbook. Before you start writing, do some research to find out about the organisms you chose. After writing your riddles, see if your classmates can solve them.

Have students work on this activity individually. Make sure that students have enough information to write their riddles. Instruct them to include the same amount of detail in their riddles as in those from the text. To help students get started, you may wish to have them find answers to the following questions about their chosen organism: Where does it live? What does it eat? How does it move? What appendages does it have? How does it produce young? What other structures does it have? Is it coldblooded or warmblooded? Students may be encouraged to select animals or plants that are unfamiliar to them so that they can practice their research skills as well as their writing skills. You may wish to supply students with the names of some organisms that they can use to write riddles about. You might even write the names of organisms on slips of paper and let students randomly choose the one they will write about. Include species of plants and animals that are common in your area. This will give students an opportunity to become more familiar with the living things around them.

Name ______________________ Date __________ Class __________

Chapter 4
Review Worksheet

Challenge Your Thinking, page 86

1. Seven Days of Diversity

In your ScienceLog, write a series of log entries that describe the different plants and animals you see during one week. Where did you see them? What were they doing? Were there any changes over the week? Did you see anything that surprised you? Compare your log entries with those of your classmates.

2. Tourist Attraction

Your city's tourist bureau has asked the members of your class to write a section in their new brochure called "Diversity in My Community." What would you write about the diversity of living things in your community in order to attract tourists to visit?

Students may highlight the variety of plants and animals in their area, focus on diversity of human life in their area, or both. Answers will vary according to region and location of the community.

3. Now You See It, Now You Don't

Create a pictorial time line that shows Krakatau before the volcano erupted and that includes the changes on Rakata up to 1985.

Students should base the ordering and content of their time line on the information provided in Lesson 1. The drawings should show the cycle of diversity among island plant and animal species. This includes the rich diversity before the explosion, the lack of diversity immediately after the explosion, and the increasing diversity in the years following the explosion.

Name ______________________ Date __________ Class __________

4. Invisible Differences

Look at the three animals pictured on page 87 of your textbook. What similarities do you see? What differences do you see? Do you think there are differences and similarities that are not visible? Explain your reasoning.

Answers will vary. A student might answer that all three animals have elongated bodies and no legs. Visible differences include the type of skin, markings on their skin, and body size. Students should realize that all three animals share common internal characteristics (blood circulation and digestion) but probably also have internal differences because of differences in their environments and food sources.

5. They're Everywhere!

What plants do you sit on? What plants do you wear, eat, write on, and sleep in? Think about the ways that plants affect your life, and list as many as you can. Describe what your life would be like if there were not so much diversity in plants.

Students should recognize the presence of many different plants and plant products all around. For example, we sit on chairs made of wood, wear clothes made of cotton, eat vegetables, write on paper made from trees, and sleep in beds that may have wooden frames and cotton sheets. Students should identify other ways that plants affect their lives, such as the production of the oxygen we need in order to survive. Each student's description of what life would be like without plant diversity should reflect an understanding of the many ways that plants are involved in his or her lifestyle.

Illustration also on page 87 of your textbook

Name ______________ Date ________ Class ________

Chapter 4 Assessment

Word Usage

1. Use all of the following terms in one or two sentences to show how they are related:

   a. diversity, organisms, biologists, differences

   Sample answer: *Biologists* use the term *diversity* to describe *differences* among *organisms*.

   b. appearance, habitat, living things, eating, characteristics, distinguish

   Sample answer: Some of the *characteristics* that *distinguish living things* from one another include *habitat, appearance,* and *eating* habits.

Correction/Completion

2. The statements below are either incorrect or incomplete. Your challenge is to make them correct and complete.

   a. The diversity of life can be seen only in certain regions of the world among specific organisms.

   The diversity of life can be seen *in every part of the world among all types of organisms.*

   b. There are more than **1.4 million** known kinds of living things in the world, but the actual number is estimated to be between **10 million** and **100 million**.

Short Response

3. Compare and contrast bats and birds.

   Sample answer: Although bats and birds may appear to have a lot in common because they are both flying animals, they are more different than they are alike. Bats are mammals and nurse their young, while birds, which are not mammals, lay eggs. Bats tend to be nocturnal, while birds tend to be most active during the day.

Name ______________ Date ________ Class ________

CHALLENGE

Illustration for Interpretation

4. Look at the illustration below. How is the diversity of life influencing these two organisms?

   Sample answer: The diversity of life is influencing the roles of these animals in a predator-prey relationship. The cheetah is a powerful hunter and has a body equipped for fast pursuit of its prey. The gazelle does not have the body type to compete or fight with the cheetah. These differences are essential to maintaining their roles in their environment.

Name ______________ Date ________ Class ________

CHALLENGE 2

Short Essay

5. Suppose that a layer of dust in the atmosphere caused a worldwide reduction in the average temperature. If this condition were to last for hundreds of years, how might the diversity of life be affected in the place where you live?

Answers will vary, but students should make it clear that all aspects of the environment—living and nonliving—are tightly interconnected. When any one aspect of the environment changes, there will be repercussions such as the extinction of a species or the evolution of a new species through adaptation. It is difficult to predict with any certainty what those repercussions will be.

Name ______________ Date ________ Class ________

Discussing the Results

1. Do these results support your hypothesis? (They supported Kettlewell's hypothesis.)

Answers will depend on the hypothesis that was developed by the students.

2. How can you explain the results?

When the light-colored bark was darkened, the light-colored moths were no longer camouflaged. They then became easy prey for predators such as birds. Under the same circumstances, however, the dark-colored moths had the advantage of camouflage and survived better.

3. Why didn't Kettlewell release the moths only into the polluted area?

The moths released into the unpolluted area were used as the control group in the experiment.

4. Do Kettlewell's findings support the theory of natural selection? Explain your answer.

Kettlewell's findings support the theory of natural selection because more of the dark-colored moths survived in the polluted area and would therefore be able to pass their dark coloration on to later generations. By contrast, the light-colored moth population decreased in the polluted area.

Name ______________ Date ________ Class ________

Darwin's Finches, continued

Questions

1. How is the structure of the beak well suited to the diet of each group of finches?

   The beak of the small tree finch is broad and short, enabling it to eat aphids and small berries. The beak of the large tree finch enables it to obtain and grind larger fruit and insects than the small tree finch is able to eat. The beak of the small ground finch is long and pointed, enabling it to obtain tiny seeds from the ground and ticks from iguanas. The large ground finch has a strong conical beak that enables it to crush hard seeds. The cactus finch has a long beak that it uses to probe for nectar in cactus flowers.

2. Do the differences in Darwin's finches support the theory of natural selection? Give reasons for your answer.

   Yes. The finches that had the appropriate beak for feeding on a particular food source had the greatest chance of surviving and passing that characteristic on to their offspring.

3. Scientists have speculated that Darwin's finches reached the Galápagos Islands from the mainland of South America as a single flock perhaps a million or more years ago. Think about the following questions, and explain your answers:

   a. What do you think the original finches looked like? Why?

   Accept all reasonable responses. (You may wish to tell students that scientists believe that the original finches probably ate seeds like those found on the South American mainland.)

   b. Is it possible that the original birds were various species that arrived on the islands at different times?

   Possibly, but probably not, because the finches have so many similarities other than their beaks

Name ______________ Date ________ Class ________

Darwin's Finches, continued

   c. Assume that one flock of finches gave rise to the 14 different species now existing on the islands. If this occurred, would it be significant that the Galápagos chain consists of many small islands rather than one large one?

   Yes. The birds' beaks adapted to the food sources available, which varied from island to island. Also, living on one large island would have allowed the finches to keep breeding with each other as a large group, and they might not have diversified.

   d. What advantages would the finches have had in arriving on the islands under the following conditions?

      i. There were no other species with exactly the same diet.

      There would have been no competition from other species for the same food.

      ii. There were no predators.

      The finch population could feed and reproduce with no threat to its existence from predators.

      iii. There were no parasites to live on the finches and weaken them.

      The finch population could remain healthy and strong without any damaging effects from parasites.

4. How has diversity helped Darwin's finches survive?

   Because of the diversity of their beaks, the finches were able to successfully inhabit many different habitats on the islands. Diversity also reduced competition for limited food sources.

Name ______________________ Date __________ Class __________

Chapter 5
Resource Worksheet

The Fine Art of Survival, page 97

Adaptations are inherited features that help increase an organism's chances of surviving and reproducing. Both animals and plants have many different kinds of structures and behaviors that help them survive in their environments. Below are three categories of adaptations to discuss: adaptations for obtaining food, adaptations for protection, and adaptations for locomotion. The organisms on pages 96 and 97 of your textbook will give you some hints.

1. Name some examples of the great variety of structures and behaviors (adaptations) that animals and plants have for obtaining their food.

Structures and adaptations for obtaining food include the Venus' flytrap's leaves (for catching insects); the spider's glands that secrete silk (to build the webs that catch its prey); the hummingbird's long, thin bill (for reaching the nectar in flowers); the plant's large leaves (for catching sunlight to make food); the lion's sharp claws and teeth (for catching and eating its prey); and the mosquito's sucking tube (to obtain blood from its victims).

2. Can you think of animal and plant adaptations that could be used for protection? Give some examples.

Structures and adaptations for protection include the cactus's sharp spines that protect it from being eaten and the blowfish's spines and ability to puff itself up as a warning to potential predators.

3. Consider an animal's locomotion—its movement from place to place. Name some adaptations that various animals have for locomotion. Is each adaptation you thought of related in some way to the organism's habitat? Explain.

Adaptations for locomotion include the long arms of the gibbon, which enable it to swing from tree branches, the wings of the hummingbird and mosquito, which enable them to fly, and the fins and tail of the blowfish, which enable it to swim.

Chapter 5

Name ______________________ Date __________ Class __________

Chapter 5
Theme Worksheet

Using the Theme of Changes Over Time

This worksheet is an extension of the theme strategy outlined on page 108 of the Annotated Teacher's Edition. It is designed to correspond to Lesson 3, The Value of Diversity, which begins on page 106 of the Pupil's Edition.

Focus question	How might human activities affect the diversity of life on Earth?

The Earth's atmosphere has changed considerably in the past 4.5 billion years. The graph below shows the changes that have occurred in the levels of gases in the atmosphere.

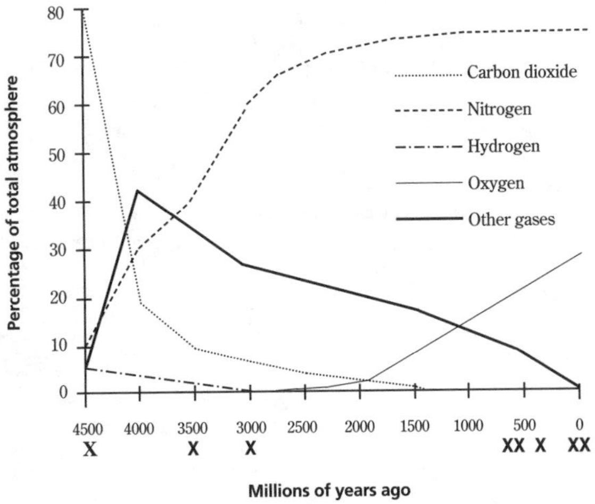

Percentage of total atmosphere

80
70
60
50
40
30
20
10
0

Carbon dioxide
Nitrogen
Hydrogen
Oxygen
Other gases

4500 4000 3500 3000 2500 2000 1500 1000 500 0
X X X XX X XX

Millions of years ago

Using the data in the chart on the next page, indicate with an "x" when a significant event took place. The first event has been marked for you.

Name ______ Date ______ Class ______

Event	Millions of years ago
Formation of the Earth	4500
First living organisms evolve in the sea.	3500
Photosynthesis begins in bacteria.	3000
First multicellular organisms evolve.	630
Plants, animals, and fungi first invade land.	410
Amphibians are the first vertebrates on land.	350
The first humans appear.	2
Tropical rain forests are being destroyed, damage to ozone layer is increasing, and pollution of atmosphere is evident.	0

Questions

1. What is remarkable about the graph at the time when photosynthesis begins in bacteria and when plants first appear?

   Oxygen levels begin to increase as photosynthesis begins in bacteria, and oxygen levels start to increase even more with the appearance of plants because oxygen is a product of photosynthesis.

2. Which gas do you think is likely to increase the most during your lifetime? Why?

   Carbon dioxide levels will probably increase the most as a growing human population develops more fossil fuel–dependent technologies and industries, which release this gas into the atmosphere.

3. What effects does damaging the atmosphere have on diversity?

   The table and the graph indicate that changes in the levels of gases in the atmosphere over time can result in huge changes in the life-forms on Earth. When oxygen became a more prominent gas, diversity increased. However, destroying rain forests, polluting the environment, and emitting dangerous chlorofluorocarbons into the atmosphere can increase levels of carbon dioxide. This could result in the endangerment and even extinction of certain species of organisms. Human activities like these may greatly affect diversity.

Name ______ Date ______ Class ______

EXPLORATION 2

Back From the Brink, page 110

Your goal to learn about factors that can lead to the extinction of a species and to determine how those factors might be avoided

The brown pelican is a fish-eating coastal bird that nests along the Atlantic, Pacific, and Gulf Coasts of the Americas. Below are some facts about brown pelicans on West Anacapa Island, a major breeding colony for brown pelicans in California. The facts trace the brown pelican's history from the days when it lived in large numbers on West Anacapa Island, through its decline, and to its comeback from the brink of extinction. The facts are not in the right sequence, however. Your task is to put them in order.

Suggested sequence:

_4_ **a.** Investigators discovered that a chemical company had been dumping DDT into the Los Angeles sewer system for some time.

_8_ **b.** An average of 5000 pairs of brown pelicans nested on West Anacapa Island from 1985 to 1989.

_7_ **c.** In 1973 the United States placed the brown pelican on its list of endangered species—those species that may not survive in the wild unless they are protected.

_3_ **d.** When brown pelicans ate DDT-contaminated fish, the DDT accumulated in their bodies. This caused the shells of their eggs to be so thin and fragile that they often broke while the eggs were being laid or incubated.

_5_ **e.** In 1970, only one brown pelican hatched on West Anacapa Island.

_1_ **f.** Brown pelicans nested in large numbers on West Anacapa Island.

_6_ **g.** In 1972 the use of DDT was banned.

_2_ **h.** DDT worked well in killing mosquitoes and other insects, but biologists discovered that DDT had contaminated the Pacific Ocean and the fish in it. This caused considerable harm to the food chain.

Name ______________ Date ________ Class ________

Exploration 2 Worksheet, continued

Another Close Call

Examine the diagram below closely. What does it tell you about the survival of the brown pelican in Texas and Louisiana? Record your observations below.

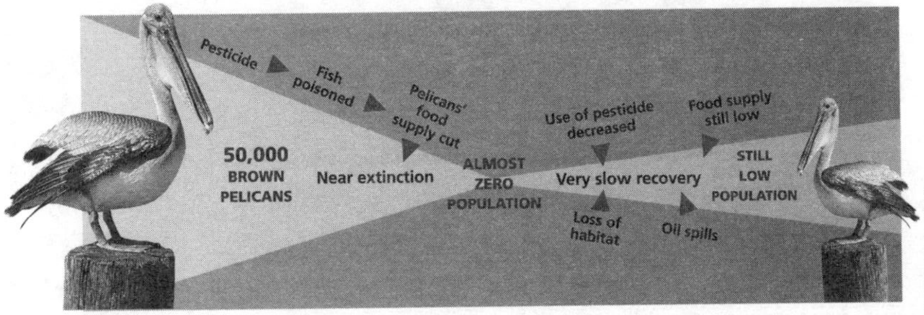

Pesticide
Fish poisoned
Pelicans' food supply cut
50,000 BROWN PELICANS
Near extinction
ALMOST ZERO POPULATION
Use of pesticide decreased
Very slow recovery
Loss of habitat
Food supply still low
Oil spills
STILL LOW POPULATION

Illustration also on page 110 of your textbook

Students' answers will vary but should indicate that in Texas and Louisiana, a pesticide poisoned the fish that brown pelicans ate. When the fish died, the pelicans' food supply was cut off, and the population of pelicans dropped to almost zero. The use of the pesticide has since decreased, but the numbers of pelicans in Texas and Louisiana will probably remain low because of decreased food supply, oil spills, and the loss of habitat.

Chapter 5

Name ______________ Date ________ Class ________

Chapter 5
Graphing Practice Worksheet

May the Best Animal Win

Complete this worksheet at the end of Chapter 5, which begins on page 88 of your textbook.

Marguerite and Susan both have pets. Marguerite's pet is a small house spider. She wants to race her spider against Susan's pet, a green turtle. Because Susan is sure her turtle is slower, she agrees to the race only if her turtle is given a 7 cm head start. Marguerite agrees. Susan recorded the results of three different races in the graphs below.

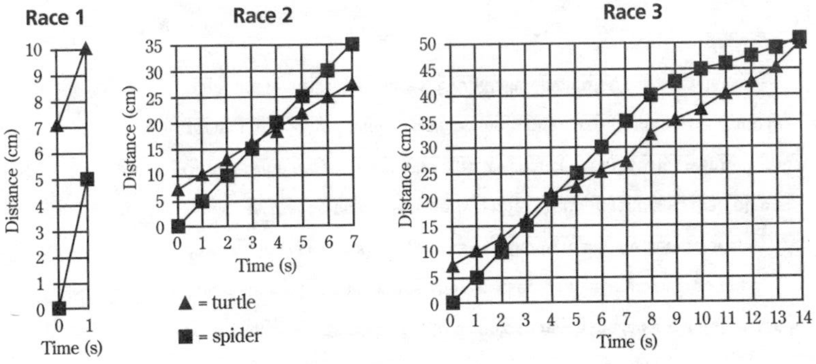

Race 1
Distance (cm)
Time (s)

Race 2
Distance (cm)
Time (s)

Race 3
Distance (cm)
Time (s)

▲ = turtle
■ = spider

Sports Commentary

1. Notice that one of the three races was 10 cm long, the next was 35 cm long, and the last was 50 cm long. Tell which animal won each race and how far ahead it was.

   The turtle wins the first race by 5 cm, the spider wins the second race by 7 cm, and the third race ends in a tie.

2. If the loser continued to walk at the same speed after the winner crossed the finish line, how long would it take for the loser to finish each race?

   In the 10 cm race, the spider walks 5 cm/s, so it would take the spider 1 second to finish. The turtle walks 3 cm/s, so it would walk its last 7 cm of the 35 cm race in $2\frac{1}{3}$ seconds.

Name ______ Date ______ Class ______

3. How do the spider's and the turtle's walking rates differ during the third race?

   Sample answer: The spider walks rapidly at first, and its speed remains at 5 cm/s for 7 seconds. It then slows down. The turtle walks at a steadier pace and is still moving at approximately the same speed when it catches up with the tired spider. That is why the longest race ends in a tie.

4. How do these differences relate to the animals' daily activities?

   Sample answer: It is likely that in daily life, the spider must move very quickly but not for great distances, an adaptation for attacking its prey. The turtle, however, is armored and is not a predator, so it may not need to walk so fast.

5. This race was held on level, dry ground. Would the results be different if it was held in water? Why?

   The turtle, unlike the spider, can swim through the water, so the turtle would always win.

6. How do these races relate to the "survival of the fittest"?

   Organisms most suited to their environments are most likely to survive. The spider won the second race because it is accustomed to walking quickly for a short period of time. The spider is not adapted for walking long distances, so the third race ended in a tie. Winning these races corresponds to surviving in a given environment.

Name ______ Date ______ Class ______

Challenge Your Thinking, page 111

1. Tall Tales

Evidence in the fossil record suggests that the ancestors of modern giraffes had very short necks. They lived on grasslands in Africa where there was a lot of vegetation that they were able to use for food. There were short grasses, bushes, and tall trees. There is also evidence of long periods of drought in that region of Africa.

How do you think Lamarck would explain the development of giraffes with long necks? How would Darwin explain it?

Sample answer: Both scientists would agree that during the long periods of drought, larger plants with deeper roots would survive longer than smaller plants with shallow roots. Both would agree that it would be advantageous for a herbivore to be able to reach the taller plants. Lamarck would suggest that when the short-necked giraffes encountered the drought, they acquired long necks in order to reach the trees. Darwin would say that some of these early giraffes had longer necks than others. Giraffes with longer necks had a better chance for survival. Over time, the giraffes with longer necks survived more often to reproduce, resulting in more giraffes with long necks. Eventually, all giraffes had long necks.

Name ____________ Date ________ Class ________

Chapter 5 Review Worksheet, continued

2. To the Editor

An article in the local newspaper stated, "Everything that humans do to the environment causes animals to become extinct." Write a letter to the editor stating whether you agree or disagree with this statement. Give several reasons for your viewpoint.

Answers will vary. Students should be able to support their opinions. Students in agreement with the statement may cite such evidence as land development and the use of pesticides. Students who disagree with the statement may cite such evidence as the protection that wildlife departments and zoos offer to endangered species. Encourage students to include in their letters their own innovative ways to prevent extinction.

3. A Different World

Imagine what the world would be like if

- all plants were 5 cm tall.
- all bears were black.
- all rabbits were white.
- no insects could fly.
- only seals lived in the ocean.

For one of these situations, list all of the ways you think the world would be different from the way it is now.

Sample responses include the following: If all plants were 5 cm tall, overcrowding would result. Plants would not be able to take advantage of the resources available in the higher and lower layers of a forest ecosystem. If all bears were black, those in the Arctic would have no camouflage. If all rabbits were white, those that live in woods and fields would easily be seen by predators. If no insects could fly, some insect-eating birds and bats would starve, and many flowers would not be pollinated. If the only living things in the oceans were seals, they would have nothing to eat.

Illustration also on page 111 of your textbook

Name ____________ Date ________ Class ________

Chapter 5 Review Worksheet, continued

4. Critter Creation

Appendages are adaptations designed to help an animal perform various specialized tasks. Your thumb is an appendage; in fact, your entire hand and arm is an appendage. Appendages help an animal live successfully in its environment.

In your ScienceLog, design and draw an animal with appendages that will give it the following characteristics:

a. The animal lives in water.

b. Its heavy body needs a lot of support as the animal walks.

c. It can walk on the bottom of a body of water for many kilometers without stopping.

d. It can dart away suddenly from its enemies by swimming.

e. It can create water currents to bring food in the water to its mouth.

f. It tests its food before eating it.

g. It has appendages that enable it to hold large pieces of food.

h. It can break apart hard bits of food.

i. Its diet includes shelled animals.

j. It has appendages that enable it to hold its young.

k. It has formidable defensive weapons.

Does the animal you drew look like any animal you have seen before? Do you think an animal exists that has all of the above characteristics? Explain your reasoning.

Students will probably be interested to learn that all of the listed features are present on a lobster or a crayfish. Lobsters have two pairs of feelers for testing food; appendages around the mouth for creating water currents; two large pincers for holding food, for breaking open shells, and for defense; four pairs of legs for walking and holding food; smaller appendages under the tail for swimming and holding eggs; and tail appendages for swimming.

Illustration also on page 112 of your textbook

Name ______________ Date ________ Class ________

Chapter 5 Assessment

Word Usage

1. Use all of the following terms in one or two sentences to show how they are related:

   a. adaptations, organisms, natural selection, reproduce, environments

   Sample answer: According to the theory of *natural selection*, *organisms* with *adaptations* that are best suited to their *environments* are more likely to survive and to *reproduce* at a higher rate than are other organisms.

   b. camouflage, mimicry, changes, snowshoe hare, avoid, hornet fly

   Sample answer: *Camouflage* helps the *snowshoe hare* survive in its changing environment because its coat *changes* color to match its surroundings. *Mimicry* protects the *hornet fly* from predators because it looks like a hornet, which predators *avoid* because hornets sting.

Correction/Completion

2. The statements below are either incorrect or incomplete. Your challenge is to make them correct and complete.

   a. Members of a species can display adaptations to a change in their environment within a short period of time.

   Members of a species may develop adaptations to a change in their environment *over a long* period of time.

Chapter 5

Name ______________ Date ________ Class ________

   b. **Jean Baptiste de Lamarck** believed that an organism could acquire a favorable characteristic during its lifetime and pass that characteristic on to its offspring.

   c. **Charles Darwin** believed that organisms with favorable characteristics were more likely to survive, reproduce, and pass those favorable characteristics to their offspring.

Short Response

3. For each adaptation on the table below, mark an "X" to show whether the adaptation helps primarily with locomotion, protection, or getting food.

Adaptation	Locomotion	Protection	Obtaining food
Eagles have extremely sharp vision.			X
Some monkeys can hold branches with their tails.	X		
Roosters have a sharp spur on each foot.		X	
Ivy plants have flat, dark green leaves.			X
Mesquite trees grow spikes on their branches.		X	
Bats have thin membranes between their fingers.	X		

CHALLENGE

Numerical Problem

4. Every spring, Nina begins to work on her lawn and gardens. First she counts the number of dandelions on her lawn. She records the number each year in order to determine how this particular area resists or succumbs to weeds. Two years ago, there were 18 dandelions, last year there were 23 dandelions, and this year there are 28 dandelions.

   a. If the dandelions keep growing at the same rate, how many dandelions will there be next year?

   There will probably be 28 + 5, or 33, dandelions.

Name ______ Date ______ Class ______

Chapter 5 Assessment, continued

b. Nina is frustrated by the greater amount of weeding she must do every year. Explain why the dandelions are so successful at reproducing.

Dandelions have developed a successful adaptation for seed dispersal. The seeds are located at the tips of fine, fringelike strands that detach easily and float away. They thrive in windy areas where even gentle breezes can send the seeds to new locations, creating new dandelions.

Chapter 5

CHALLENGE 2

Short Essay

5. Applying the theory of natural selection, explain how the ancestors of the air-breathing porpoise, who were land animals, developed the ability to live in the water.

Sample answer: Over time, the environment that the ancestors of the air-breathing porpoises lived in changed, requiring the species to live near or in the water. A few ancestors had characteristics that were favorable to living in the ocean. These ancestors were selected by their environment for survival. They could successfully reproduce, and their offspring also survived to reproduce because they inherited the characteristics that allowed them to adapt successfully to their environment. Over many generations the porpoises evolved to become ocean dwellers.

Name ______ Date ______ Class ______

Chapter 6 Resource Worksheet

A Simplified Classification System for Invertebrates, page 123

Look at the animals pictured on pages 120 and 121 of your textbook. Fill in each blank in the diagram below with an animal's name.

Invertebrates

Annelids	Mollusks	Echinoderms	Arthropods	Other invertebrates*
earthworm	coquina clam	spiny sun sea star		helmet jellyfish
leech	oysters	sand dollar		horned flatworm
	scallop	sea urchin		purple veined anemone
	octopus			rotifer

Arthropods

Insects	Arachnids	Crustaceans	Diplopods/chilopods
dogbane leaf beetle	Argiope spider	crayfish	striped millipede
grasshopper	harvestman daddy longlegs	jonah crab	centipede
painted lady butterfly	tarantula		

*"Other invertebrates" includes other worms, such as flatworms and unsegmented roundworms, along with many water animals, such as jellyfish, sea anemones, corals, and sponges.

Name ______________ Date ________ Class ________

A Simplified Classification System, continued

Questions

Now think some more about the subgroups.

1. Did you have any problems deciding which subgroup each invertebrate belonged to? If so, which one(s)?

Students will probably have some problems classifying invertebrates. It is difficult to classify invertebrates using only photographs, especially if they are not familiar with the organisms' structures.

2. It's interesting to think about where invertebrates live. How many are found in water?

Most of these animals live in the water.

in moist places?

Earthworms, snails, centipedes, and millipedes prefer moist land environments.

on dry land?

Insects and arachnids can live on dry land.

3. Does it appear that the structure of invertebrates enables them to live successfully in various places? Why do you think this is the case?

Yes. Because of the diversity of structures among invertebrates, they can live successfully in many different places.

Chapter 6

Name ______________ Date ________ Class ________

Chapter 6
Resource Worksheet

Putting It All Together, page 126

Many animals are shown on pages 126–129 of your textbook. Your task is to classify them using the classification tables below. These tables are an overall classification system for animals. They bring together everything you have learned about classification for the animal kingdom.

If your classroom contains any living or preserved specimens that are not represented in the pictures, classify them as well.

Animal Kingdom Classification System

Invertebrate subgroups		Examples
Annelids		**(j) bearded bristleworm, (u) leech**
Mollusks		**(b) tree snail, (h) clam**
Echinoderms		**(v) starfishes, (r) brittle star**
Arthropods	Insects	**(f) soldier beetle**
	Crustaceans	**(d) Sally lightfoot crab**
	Arachnids	**(a) scorpion**
	Diplopods/chilopods	**(g) millipede**
Other invertebrates		**There are no other invertebrates.**

Vertebrate subgroups	Examples
Fishes	**(q) goldfish**
Amphibians	**(s) dwarf American toad, (t) blue-spotted salamander**
Reptiles	**(n) green sea turtle, (k) tree boa**
Birds	**(e) white-faced tree ducks, (m) Adélie penguins, (o) ostriches**
Mammals	**(c) manatees, (i) Arabian horses, (l) red squirrel, (p) bottlenose dolphins**

Name ______ Date ______ Class ______

Exploration 1 Worksheet, continued

Analyze Your Data

1. Study your table and compare your results with those of the other groups in your class.
2. Calculate the percentage of individuals in your group who have each characteristic. What does this tell you?

   Answers will vary. As students complete the table and answer the questions, they will realize that there are hundreds of characteristics within the human species that can vary.

3. Does anyone have all of characteristics 1, 3, 5, 7, 9, and 11? Does anyone have all of characteristics 2, 4, 6, 8, 10, and 12?

   Answers will vary.

   Are there any two people in the class who have identical characteristics? Did you know that there are 64 possible combinations of these six characteristics?

   Answers will vary.

4. You have looked at only six characteristics. There are hundreds of other characteristics you might have considered. What are some of these other characteristics?

   Other examples include tooth size and shape, hair texture, and presence of freckles. Students should conclude that no two people are exactly alike.

Name ______ Date ______ Class ______

Chapter 6
Review Worksheet

Challenge Your Thinking, page 136

1. The Animal-Kingdom Pie Can you explain the meaning of this pie chart? Which animals would you place in the missing piece of pie?

The whole pie represents all of the animal kingdom, three-quarters of which are arthropods. Of the arthropods, the insects represent the largest section. The missing portion contains all other invertebrate subgroups, but primarily it contains vertebrates, or animals with backbones.

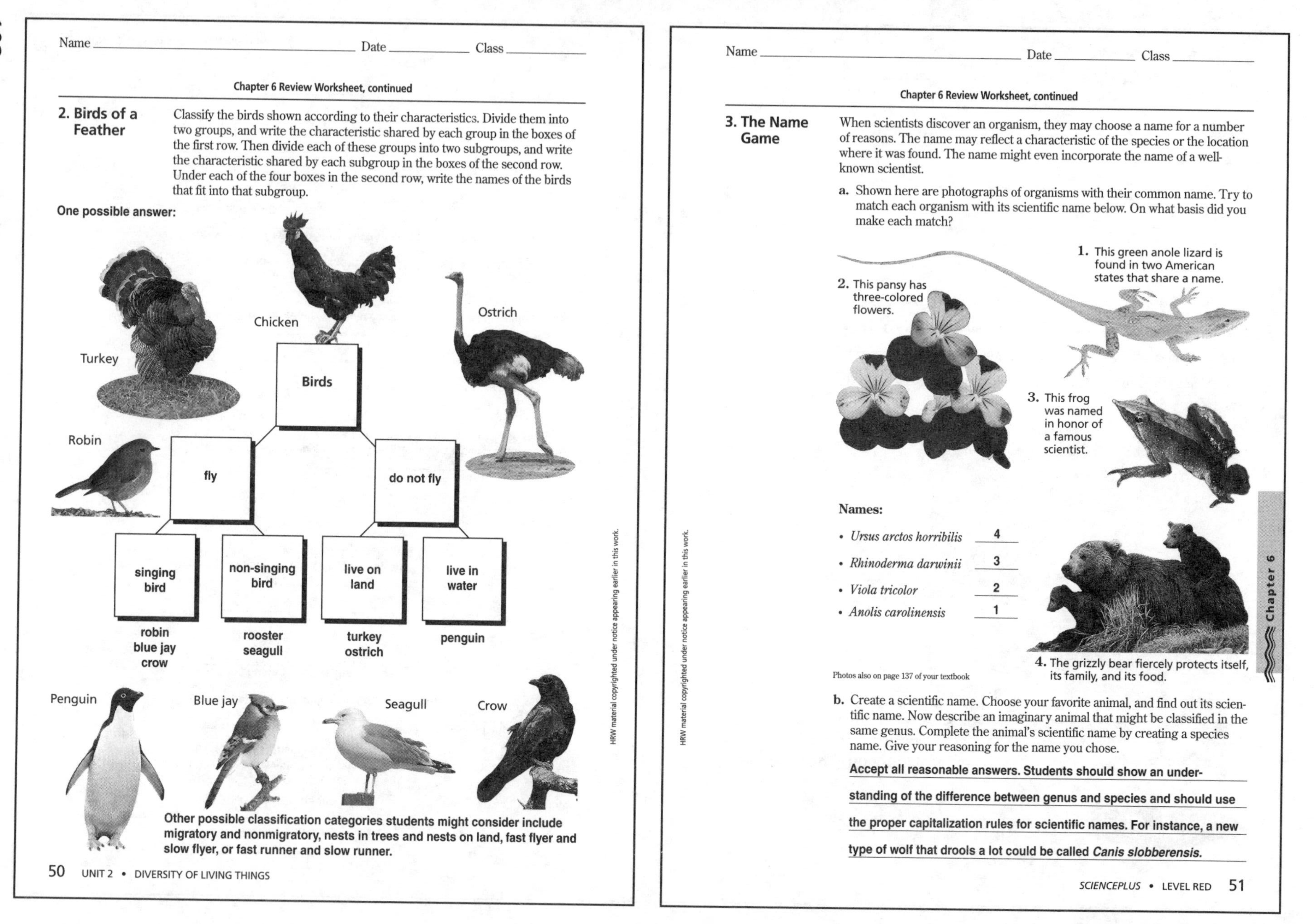

Name ______________________ Date __________ Class __________

Chapter 6 Review Worksheet, continued

2. Birds of a Feather

Classify the birds shown according to their characteristics. Divide them into two groups, and write the characteristic shared by each group in the boxes of the first row. Then divide each of these groups into two subgroups, and write the characteristic shared by each subgroup in the boxes of the second row. Under each of the four boxes in the second row, write the names of the birds that fit into that subgroup.

One possible answer:

Other possible classification categories students might consider include migratory and nonmigratory, nests in trees and nests on land, fast flyer and slow flyer, or fast runner and slow runner.

Name ______________________ Date __________ Class __________

Chapter 6 Review Worksheet, continued

3. The Name Game

When scientists discover an organism, they may choose a name for a number of reasons. The name may reflect a characteristic of the species or the location where it was found. The name might even incorporate the name of a well-known scientist.

a. Shown here are photographs of organisms with their common name. Try to match each organism with its scientific name below. On what basis did you make each match?

1. This green anole lizard is found in two American states that share a name.

2. This pansy has three-colored flowers.

3. This frog was named in honor of a famous scientist.

4. The grizzly bear fiercely protects itself, its family, and its food.

Names:

- *Ursus arctos horribilis* **4**
- *Rhinoderma darwinii* **3**
- *Viola tricolor* **2**
- *Anolis carolinensis* **1**

Photos also on page 137 of your textbook

b. Create a scientific name. Choose your favorite animal, and find out its scientific name. Now describe an imaginary animal that might be classified in the same genus. Complete the animal's scientific name by creating a species name. Give your reasoning for the name you chose.

Accept all reasonable answers. Students should show an understanding of the difference between genus and species and should use the proper capitalization rules for scientific names. For instance, a new type of wolf that drools a lot could be called *Canis slobberensis*.

Name ______________ Date ________ Class ________

4. Can You Relate?

According to the classification systems used by biologists, why is

a. a turtle more closely related to an alligator than to an eel?

A turtle is more closely related to an alligator than to an eel because both the turtle and the alligator are reptiles, while the eel is a fish.

b. an earthworm more closely related to a grasshopper than to a snake?

An earthworm is more closely related to a grasshopper than to a snake because both the earthworm and the grasshopper are invertebrates, while the snake is a vertebrate.

c. an octopus more closely related to a clam than to a lobster?

An octopus is more closely related to a clam than to a lobster because both the octopus and the clam are mollusks, while the lobster is a crustacean.

d. a whale more closely related to a tiger than to a shark?

A whale is more closely related to a tiger than to a shark because both the whale and the tiger are mammals, while the shark is a fish.

e. a maple tree more closely related to a rosebush than to a pine tree?

A maple tree is more closely related to a rosebush than to a pine tree because both the maple tree and the rosebush are flower-bearing plants, while the pine tree is a cone-bearing plant.

Name ______________ Date ________ Class ________

Word Usage

1. Use all of the following terms in one or two sentences to show how they are related:

a. animal, subgroup, Linnaeus, vertebrates, kingdom, classification, group

Sample answer: The largest *group* of living things in *Linnaeus's* system of *classification* is the *kingdom*. One *subgroup* of the *animal* kingdom is the *vertebrates*—animals with a backbone.

b. arthropods, lobster, invertebrates, subgroup, kingdom

Sample answer: A *lobster* is a member of the *subgroup* crustaceans, which is a subgroup of *arthropods*, which is a subgroup of *invertebrates*, which is a subgroup of the animal *kingdom*.

Correction/ Completion

2. The statements below are either incorrect or incomplete. Your challenge is to make them correct and complete.

a. *Homo sapiens* is the scientific name for **human beings**.

b. According to modern biologists, mushrooms, mosses, and maples all belong to the plant kingdom.

According to modern biologists, mosses and maples belong to the plant kingdom, *but mushrooms belong to the other (fungi) kingdom.*

Short Responses

3. For each animal below, write either *V* for vertebrate or *I* for invertebrate.

V frog **V** shark **I** mosquito
I lobster **V** bat **I** snail

Name ______________ Date ________ Class ________

CHALLENGE 1

Illustration for Interpretation

4. Circle the two organisms in each group that are most closely related to each other. Then tell what important characteristic they have in common.

a. (wasp) (earthworm) eel — **Both are invertebrates.**

b. (seal) tuna (horse) — **Both are mammals.**

c. (rosebush) (oak) fern — **Both are flowering plants.**

d. robin (bat) (armadillo) — **Both are mammals.**

5. Look at the marine environment below and then use the space that follows to make a classification diagram for the organisms shown.

Sample answer:

- **Living things**
  - **plants**
    - **land plants**
    - **sea plants**
  - **animals**
    - **vertebrates**
    - **invertebrates**

Name ______________ Date ________ Class ________

CHALLENGE 2

Short Essay

6. Why do scientists find it useful to have a consistent system for classifying living things?

Sample answer: By using a consistent classification system, scientists not only keep track of millions of different organisms but also learn more about them. Using one language to name organisms eliminates confusion among international scientists when referring to a specific organism. Since the classification system requires studying each organism's interior as well as its exterior, scientists learn how organisms carry on their life functions and how they are related to other organisms. For example, whales look much like fish, but because they have lungs and mammary glands, they are actually more closely related to humans.

Name ______________________ Date ____________ Class ____________

Unit 2
Unit Activity Worksheet

Test Your Memory

Use this activity as you conclude Unit 2.

Search through Unit 2 to locate each word for the given definition. The numbers under the blanks will help you complete the poem on the next page.

a. soft-bodied animals, most of them having shells: M O L L U S K S (K = 21)

b. animals with feathers: B I R D S (D = 1)

c. most mammals are born live, but reptiles, birds, fish, and amphibians develop from: E G G S (G = 14)

d. simple land plants: M O S S E S (E = 4)

e. a group of animals or plants that have certain permanent characteristics in common:
S P E C I E S (C = 19)

f. adaptations are what allow an organism to: S U R V I V E (R = 5)

g. animals without backbones: I N V E R T E B R A T E S (B = 16)

h. the Swedish scientist who devised a scientific way of classifying living things by their similarities:
L I N N A E U S (N = 10)

i. animals in this subgroup of arthropods have six legs and wings: I N S E C T S (I = 2)

j. an animal with soft damp skin: A M P H I B I A N (P = 18)

k. animals with backbones: V E R T E B R A T E S (V = 3)

l. young grasshoppers whose diet determines their color: N Y M P H S (Y = 8)

m. worms with many segments: A N N E L I D S (L = 10)

n. the scientific name for human beings: H O M O S A P I E N S (P = 18)

o. animals with scales and fins: F I S H (F = 13, H = 15)

p. the person responsible for the theory of natural selection: C H A R L E S (S = 6)
D A R W I N

q. animals with jointed feet or legs: A R T H R O P O D S (O = 9)

r. features of organisms that enable them to survive and reproduce:
A D A P T A T I O N S (T = 7)

Name ______________________ Date ____________ Class ____________

Test Your Memory, continued

s. an imitative, attention-getting behavior: M I M I C R Y (M = 17)

t. Darwin's explanation of how the features of a species can change over many generations:
N A T U R A L S E L E C T I O N (U = 20, L = 11)

u. animals with fur or hair: M A M M A L S (A = 12)

v. animals with scales but no fins: R E P T I L E S (T = 7)

A Poem

Using the appropriate letters for the code numbers from the definitions above and on the previous page, fill in the blanks to discover a short poem.

D I V E R S I T Y
1 2 3 4 5 6 2 7 8

D I V E R S I T Y O N L A N D, O N
1 2 3 4 5 6 2 7 8 9 10 11 12 10 1 2 10

A I R, I N S E A,
12 2 5 2 10 6 4 12

D I V E R S I T Y O F L I V I N G
1 2 3 4 5 6 2 7 8 9 13 11 2 3 2 10 14

T H I N G S — A F L Y, A B I R D,
7 15 2 10 14 6 12 13 11 8 12 16 2 5 1

A T R E E,
12 7 5 4 4

D I V E R S I T Y O F F O R M, O F
1 2 3 4 5 6 2 7 8 9 13 13 9 5 17 9 13

S H A P E, O F S I Z E,
6 15 12 18 4 9 13 6 2 4

T H A T ' S W H A T M A K E S
7 15 12 7 6 15 12 7 17 12 21 4 6

E A C H L I V I N G T H I N G
4 12 19 15 11 2 3 2 10 14 7 15 2 10 14

S U C H A G R E A T S U R P R I S E!
6 20 19 15 12 14 5 4 12 7 6 20 5 18 5 2 6 4

Name ______________ Date ______ Class ______

Unit 2
Unit Review Worksheet

Making Connections, page 138

The Big Ideas

In your ScienceLog, write a summary of this unit, using the following questions as a guide:

1. What is diversity? (Ch. 4)
2. Why is there diversity among living things? (Ch. 4)
3. What are some adaptations animals have for protection? for obtaining food? for attracting a mate? for locomotion? (Ch. 5)
4. How are plants adapted to survive and reproduce? (Ch. 5)
5. What is meant by natural selection? (Ch. 5)
6. What evidence did Darwin use to develop his theory of how life evolved? (Ch. 5)
7. What kinds of conditions cause species to become extinct? (Ch. 5)
8. Why do we classify things? (Ch. 6)
9. How, in a general way, are living things classified? (Ch. 6)
10. How would you classify yourself, according to Linnaeus's system? (Ch. 6)
11. What is the scientific name of your species? (Ch. 6)

A sample unit summary is provided on page 138 of the Annotated Teacher's Edition.

Checking Your Understanding

1. For each group of words below, write one or two sentences that show how the words are related to each other.

   a. adaptation, predator, peppered moths, habitat

   Sample answer: *Peppered moths* use camouflage as an *adaptation* that protects them from *predators* in their *habitat*.

   b. organisms, diversity, environment, survival

   Sample answer: Adaptations help *organisms* in their struggle for *survival* in a given *environment*. *Diversity* arises from variations in how each species adapts over time to its habitat.

Name ______________ Date ______ Class ______

Unit 2 Review Worksheet, continued

   c. diversity, species, inherited characteristics, unique

   Sample answer: While *inherited characteristics* make members of a *species* alike in many ways, there is also great *diversity* within a species. People belonging to the same family, for instance, may look similar, but despite family resemblance, every human is *unique*.

2. Classify the living things in the photos on page 139 of your textbook according to some consistent system. Explain the system you used.
   Sample answer:

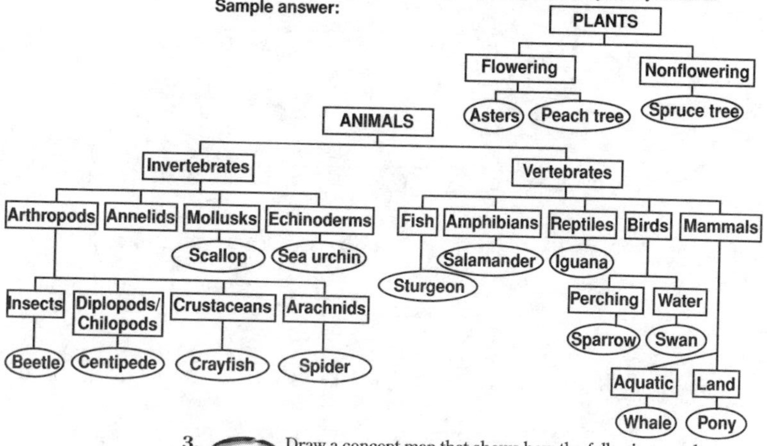

- PLANTS
  - Flowering: Asters, Peach tree
  - Nonflowering: Spruce tree
- ANIMALS
  - Invertebrates
    - Arthropods
      - Insects: Beetle
      - Diplopods/Chilopods: Centipede
      - Crustaceans: Crayfish
      - Arachnids: Spider
    - Annelids
    - Mollusks: Scallop
    - Echinoderms: Sea urchin
  - Vertebrates
    - Fish: Sturgeon
    - Amphibians: Salamander
    - Reptiles: Iguana
    - Birds
      - Perching: Sparrow
      - Water: Swan
    - Mammals
      - Aquatic: Whale
      - Land: Pony

3. (concept map) Draw a concept map that shows how the following words are related to each other: coral snake; camouflage; mimicry; snowshoe hare; adaptations; snow; grasses, trees, and weeds; brown in summer; white in winter; and king snake.
   Sample concept map:

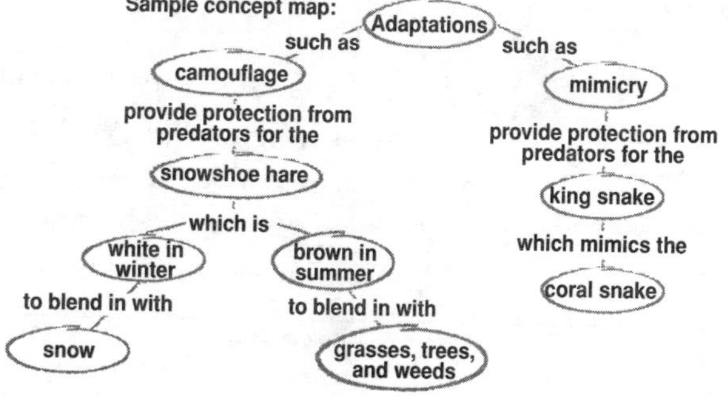

   Adaptations — such as — camouflage — provide protection from predators for the — snowshoe hare — which is — white in winter — to blend in with — snow; which is — brown in summer — to blend in with — grasses, trees, and weeds

   Adaptations — such as — mimicry — provide protection from predators for the — king snake — which mimics the — coral snake

Name ____________ Date ________ Class ________

Unit 2
End-of-Unit Assessment

Word Usage

1. Use all of the following terms in one or two sentences to show how they are related.
   a. Charles Darwin, finches, natural selection, adaptations
   Sample answer: *Charles Darwin* studied *adaptations* in *finches* when forming his theory of *natural selection.*

   b. attract, hornet, mate, sting, peacock, adaptations, predators
   Sample answer: Some animals have *adaptations* that help them *attract* attention. A male *peacock* displays his feathers to tempt his *mate,* while the black-and-yellow pattern of a *hornet* warns enemies and *predators* of its powerful *sting.*

2. The woolly mammoth was an elephantlike animal with a heavy coat. It lived in North America about 10,000 years ago, until the end of the last ice age. Use the terms below to make a hypothesis about the disappearance of these animals.
   extinct, environmental change, natural selection, adapt
   Sample hypothesis: After the last ice age, there were great *environmental changes* as the climate became warmer and the icecaps melted. According to the theory of *natural selection,* those species that could not *adapt* to the warmer climate, like the woolly mammoth, were not selected for survival and became *extinct.*

Name ____________ Date ________ Class ________

Correction/ Completion

3. The statements below are either incorrect or incomplete. Your challenge is to make them correct and complete.
   a. Arthropods are successful vertebrates because they survive in many different habitats.
   Arthropods are successful *invertebrates* because they survive in many different habitats.

   b. Scientists use a **classification** system to show the evolutionary relationship of organisms.

Short Responses

4. Using an example, describe an adaptation that a plant or animal has for the following:
   a. "disappearing" into its environment
   Sample answer: Camouflage enables an organism to "disappear" into its surroundings. Examples will vary. The walking stick is an insect that looks so much like a stick that it often goes undetected by predators.

   b. storing water
   Sample answer: Specialized parts of plant cells enable plants to store a lot of water. Cactuses store water in thickened, fleshy stems that have reduced or spinelike leaves and a heavy, waxy coating that prevents water loss.

Name ______________________ Date ____________ Class ____________

5. Seeds can be transported by wind, water, animals, or mechanical propulsion (like a bullet). Describe an adaptation that a seed might have for transportation by one of these methods.

Sample answers: Seeds that have wings or tufts could be carried by wind. Seeds that have waterproof air sacs could float on and be transported by water. Seeds that have a tough coat that resists digestion could be dispersed by means of animal feces. Seeds positioned on a high, flexible stalk could be thrown from the plant and transported by mechanical propulsion.

6. How might Darwin have used the following observations to support his theory of natural selection?

   a. Finches that live in different parts of the Galápagos Islands have different kinds of beaks.

   Finches living in different parts of the Galápagos Islands have similar features of a common ancestor but have evolved different types of beaks to eat the food available in different areas of their environments. Only those individuals with beak adaptations that allowed them to eat the available food survived to reproduce.

   b. Fossilized seashells are found high in the Andes Mountains of South America.

   The fact that fossilized seashells are found in the Andes Mountains shows how drastically an environment can change. An area that once was populated by ocean-living organisms changed over time to become mountainous terrain. Organisms must adapt to a changing environment in order to survive.

Name ______________________ Date ____________ Class ____________

Data for Interpretation

7. In the table below are some characteristics for students in an eighth-grade class. Use the table to answer the questions that follow.

Characteristic		Van	Tasha	Tiffany	Bo	Josh
Hand folding, thumb position	left over right		X		X	X
	right over left	X		X		
Ear lobes	attached	X	X			X
	free			X	X	
Hairline	pointed		X		X	X
	straight	X		X		
Tongue rolling	can roll	X	X			
	can't roll			X	X	X

   a. How many have three identical characteristics? Who are they?

   Two pairs of students: Tasha and Josh, and Bo and Josh

   b. How many have only two identical characteristics? Who are they?

   Four pairs of students: Van and Tasha, Van and Tiffany, Tasha and Bo, and Tiffany and Bo

   c. How many have only one identical characteristic? Who are they?

   Two pairs of students: Van and Josh, and Tiffany and Josh

   d. Are there any two people who are different in every respect? If so, who are they?

   Two pairs of students: Van and Bo, and Tasha and Tiffany

Name ________________ Date ________ Class ________

Unit 2 Assessment, continued

e. How does this data relate to diversity?

Sample answer: The diversity of living things is due to the many different combinations of characteristics within a species.

Graph for Interpretation

8. The graph below estimates the number of named animal species in the world. Use the graph to answer the questions that follow.

Number of animal species

1,000,000
900,000
800,000
700,000
600,000
500,000
400,000
300,000
200,000
100,000

Annelids Mollusks Arthropods Vertebrates All others

a. What is the total number of different kinds of animals?

There are almost 1.2 million named species of animals in the world.

b. Which animals are in the majority?

Arthropods are in the majority, with almost 1 million different species.

c. There are about **five** times as many arthropods in the world as all other animals put together.

Name ________________ Date ________ Class ________

Unit 2 Assessment, continued

d. What can you conclude about the diversity of the animals in this graph?

The number of different animals in each group shows the diversity of the group. Arthropods are the most diverse because there are so many species. These organisms have the ability to adapt to their environments and reproduce successfully. The other animals shown in the graph do not have the level of diversity that the arthropods do.

CHALLENGE

Numerical Problem

9. Suppose that the diversity of life on Earth were measured by percentages and that the 1.4 million known species correspond to a diversity level of 100 percent.

a. What does this mean?

Having a diversity level of 100 percent suggests that all habitats and niches are filled by the 1.4 million known existing organisms.

b. There are 250,000 species of angiosperms, or flowering plants. What percentage of the known existing species on Earth do angiosperms make up?

$\frac{250,000}{1,400,000} = \frac{17.9}{100}$ **Angiosperms make up about 17.9 percent of the known existing species on Earth.**

c. How would the destruction of all angiosperms affect the diversity level of life on Earth?

The diversity measurement would change from 100 percent to 100 – 17.9, or 82.1, percent. It is possible that the loss of angiosperms would mean the loss of other, dependent, species. According to the theory of natural selection, however, all habitats and niches made available by these losses would eventually be filled by other organisms, and the diversity measurement would eventually return to 100 percent.

Name ______ Date ______ Class ______

Illustration for Interpretation

10. Describe how each of these animals is adapted to survive in its environment. Mention adaptations for obtaining food, for protection, or for locomotion.

a. b. c.

a. **Since the rabbit is a herbivore, it must have adaptations for safe grazing, or grazing without being attacked by predators. It has peripheral vision and long ears that enable it to see and hear approaching predators. It also has strong leg muscles that allow it to move quickly to escape an attack by a predator.**

b. **The chimpanzee eats mainly fruits, leaves, and seeds and uses its hands to hold tools to search for termites and ants. The long arms of a chimpanzee enable it to climb from tree to tree to escape predators as well as to nest in the branches.**

c. **The mountain lion has large paws, sharp teeth, and speed to capture its prey. It runs quickly because it has strong leg muscles. The mountain lion has no natural predators.**

Name ______ Date ______ Class ______

CHALLENGE 2

Short Essay

11. What are some of the ways in which you benefit from the diversity of life?

Answers will vary, but students should demonstrate an understanding of the amount of diversity even in their local environments. They may wish to focus on the interdependence of organisms in terms of food acquisition, or they may address the balance that exists between various organisms in nature.

Illustration for Interpretation

12. Examine the living things pictured below, and answer the questions that follow.

Sunflower, Butterfly, Cow, Rosebush

Earthworm, Pine tree with cones, Ostrich, Fern

a. On the next page, create a chart to classify these organisms into groups and subgroups. Then divide each of the subgroups so that each organism has its own subgroup. Explain the characteristic that distinguishes each organism in your classification system.

Name ______ Date ______ Class ______

Unit 2 Assessment, continued

Your Classification Chart:

Sample answer:

Plants

sunflower
fern
pine tree with cones
rosebush

- sunflower, rosebush (with flowers)
  - sunflower (no thorns)
  - rosebush (thorns)
- fern, pine tree (no flowers)
  - fern (no cones)
  - pine tree (cones)

Animals

earthworm
ostrich
butterfly
cow

- ostrich, cow (vertebrates)
  - ostrich (bird)
  - cow (mammal)
- earthworm, butterfly (invertebrates)
  - earthworm (no wings)
  - butterfly (wings)

b. According to classification systems used by biologists today, why is a butterfly more closely related to an earthworm than to an ostrich?

Sample answer: A butterfly is more closely related to an earthworm than to an ostrich because both the butterfly and the earthworm are invertebrates, while the ostrich is a vertebrate.

Name ______ Date ______ Class ______

SourceBook Activity Worksheet, continued

Mouse and habitat	Adaptations	Environmental conditions
Brown or gray mice at the base of the mountain	• Eat a wide variety of foods, including nuts, grasses, leaves, insects, berries, and food left behind by people • Short hair	• Abundant and varied food sources • Warm temperatures
Light tan, mottled brown mice halfway up the mountain (3000 m)	**• Medium-sized ears that prevent loss of heat** **• Nest at base of trees and among fallen leaves, needles, and cones**	**• Cool temperatures** **• Fir trees provide shelter.**
White mice at the top of the mountain (5500 m)	**• White fur for camouflage** **• Eat bark and buried insects**	**• Snow on the ground** **• Limited food sources**

Conclusions

Responses will probably range from very simple to very complex. Accept all responses which indicate that the student understands Darwin's theory of evolution by natural selection as described in the SourceBook.

Name ______________________ Date __________ Class __________

Unit 2 SourceBook Review Worksheet

SourceBook

Unit CheckUp, page S49

Concept Mapping

The concept map below illustrates major ideas in this unit. Complete the map by supplying the missing terms. Then extend your map by answering the additional question below.

Sample concept map:

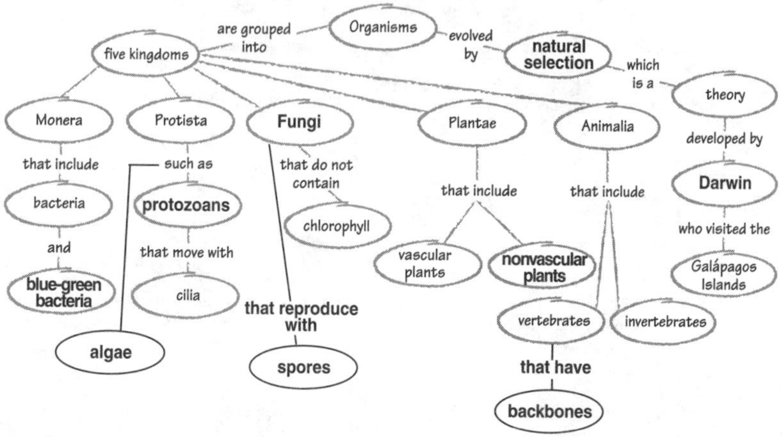

Where and how would you connect the terms *spores, algae,* and *backbones?*

Name ______________________ Date __________ Class __________

SourceBook Review Worksheet, continued

Checking Your Understanding

Select the choice that most completely and correctly answers each of the following questions.

1. Which is NOT a kingdom name?
   - a. Protista
   - b. Animalia
   - c. (Protozoa)
   - d. Fungi
2. An animal with a backbone is called
   - a. a segmented worm.
   - b. a protist.
   - c. an echinoderm.
   - d. (a vertebrate.)
3. Which is a component of fungi but NOT of plants?
   - a. chlorophyll
   - b. (hyphae)
   - c. a nucleus
   - d. a cell wall
4. During what interval did most of the Earth's history occur?
   - a. Cambrian era
   - b. (Precambrian time)
   - c. Mesozoic era
   - d. Paleozoic era
5. Similarly constructed limbs on different vertebrates are called
   - a. (homologous structures.)
   - b. biochemical similarities.
   - c. wings.
   - d. mutational structures.

Interpreting Graphs

Bacteria are grown in a test tube that contains a limited amount of food. This graph illustrates how the population of bacteria changes over a two-day period. What do you predict will happen to the size of the population after day two? Support your prediction with an explanation on the next page.

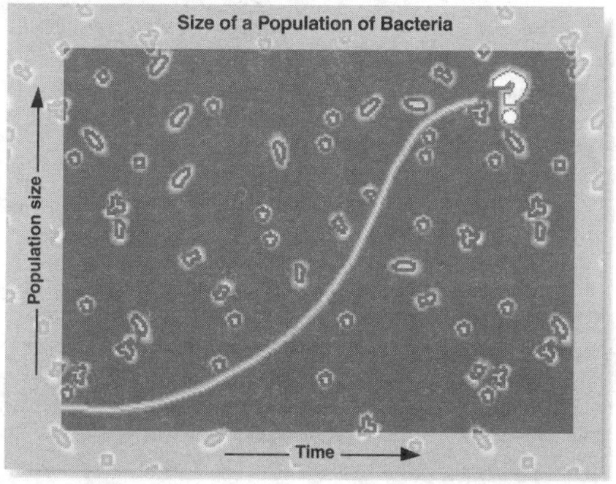

Graph also on page S50 of your textbook

Name ______________________ Date ____________ Class ____________

After day two, the size of the population will probably stabilize for a short while. But because there is a limited food supply, the struggle for existence will soon cause the population to start decreasing. Eventually the food will be used up and the entire population will die out.

Critical Thinking

Carefully consider the following questions, and write a response in the space below that indicates your understanding of science.

1. A biologist notices a large number of short plants growing near a swampy area. Although she thinks they are nonvascular plants, she is not sure. She collects one of the plants to bring back to her laboratory, where she will look at a slice from its stem under a microscope. Why does she wish to observe the plant's stem? Why is her guess that the plants are nonvascular a logical assumption?

   Looking at the stem under a microscope will reveal if the stem contains vascular tissue. Nonvascular plants do not grow tall and they must live in moist environments. It makes sense to assume that the plants in question are nonvascular—they are short and grow near a swampy area.

2. Explain why an earthworm is able to crawl faster than a planarian can move.

   A planarian is a flatworm. Flatworms have a simple nervous system. Earthworms are segmented worms that have a complex nervous system consisting of a simple brain and a nerve cord. The more advanced nervous system of the earthworm allows this animal to crawl faster than the planarian can.

Name ______________________ Date ____________ Class ____________

3. There are many species of small birds called finches that live on the Galápagos Islands. Scientists think that the many different kinds evolved from a single species of finch that flew to the islands years ago. Is it more likely that this ancestral species arrived from Europe, North America, or South America? Explain the reason for your choice.

   South America, because the Galápagos Islands are near South America. The other two places are much farther away.

4. You discover a new species of protozoan but are unsure where to place it on the protist evolutionary tree. What kind of observations would you make about the protist, and what kind of analyses would you perform on it to help you classify this organism?

   Look at the protozoan under a microscope and see if it moves around. If it is able to move, then try to determine *how* it moves (for example, by flowing, like an amoeba, or with cilia, like a paramecium). Determining how it moves will indicate which protozoan group it belongs to. If it doesn't move, it may be some sort of parasitic protozoan. To further pinpoint the classification of this organism, you should conduct a DNA analysis and compare its DNA with the DNA of other, known protozoans.

5. The wing of a bird and the wing of an insect have the same function, yet they are NOT homologous structures. Explain why they are not.

   They are not homologous because their structure is different.

Portfolio Idea

In your ScienceLog, create a table that summarizes the major characteristics of the organisms in each of the five kingdoms. Start by placing the kingdom names as headings in your table. Then compare and contrast the organisms in each kingdom to come up with the characteristics you will have in your table. Make your table large enough to include a colored drawing or a magazine clipping of a representative organism for each kingdom.

Name ______________________ Date __________ Class __________

Unit 2 SourceBook Assessment

SourceBook

1. Fill in the blanks with the following subcategories, in order, beginning with kingdom and ending with the most specific classification: class, family, species, phylum.

   kingdom, **phylum**, **class**, order, **family**, genus, **species**

2. Put the following terms in the proper order, from the longest period of time to the shortest: era, decade, period, week, year.

   era, period, decade, year, week

3. Monerans, protists, animals, and fungi are examples of the category
   a. phylum. b. class. c. species. d. (kingdom.)

4. Cyanobacteria contain chlorophyll and produce their own food by a process called
   a. respiration. b. (photosynthesis.) c. conduction. d. osmosis.

5. Protozoans are animal-like protists; therefore, they (do/(do not)) have a cell wall, and they (do/(do not)) contain chlorophyll. (Circle the correct words.)

6. The black or gray powdery substance that you may find growing on an old hamburger bun is an example of a(n)
   a. (fungus.) b. cyanobacterium. c. alga. d. moneran.

7. Redwood trees are
   a. (vascular plants.) b. nonvascular plants.

8. The most diverse group of modern plants are those with
   a. "naked" seeds. b. ("covered" seeds.)

9. Which of the following does NOT correctly complete the sentence? You are
   a. (an invertebrate.) b. an animal. c. warmblooded. d. a mammal.

10. Animals must
    a. be multicellular. b. have organ systems. c. have vertebrae.
    d. be able to make their own food within their body.
    e. All of the above f. (a and b only) g. a, b, and d only

11. A man walked into a restaurant and ordered an arthropod. Which of the following did he order?
    a. steak b. catfish c. (lobster) d. chicken

Name ______________________ Date __________ Class __________

12. A person's bones are called an
    a. (endoskeleton.) b. exoskeleton.

13. Coldblooded refers to
    a. (animals having a body temperature that changes according to the temperature of the environment.)
    b. all vertebrates that live in the water.
    c. animals having a body temperature below 20°C.
    d. **a** and **b** only

14. Which invertebrates have gills when they are young and lungs when they mature? These animals also live part of their lives in water and part on land.
    a. reptiles b. coelenterates
    c. (amphibians) d. segmented worms

15. The term ***millennium*** means
    a. 10,000 years. b. 5 decades.
    c. any long period of time. d. (1000 years.)

16. A blue jay laid two eggs. The first egg hatched a very strong and healthy female blue jay. The second egg hatched a much weaker and smaller female blue jay. The first blue jay was able to learn to fly, mate, and reproduce. The second was unable to fly very well and died in just a few weeks. What process is described here?
    a. fate b. (natural selection) c. adaptation d. mutation

17. Farmer Jane grew tomatoes. She took the seeds from the largest, reddest, and best-tasting tomatoes and planted them. From the plants that grew, she again took the seeds from the largest, reddest, and best-tasting tomatoes and planted them. She repeated this for three seasons. Farmer Jane then had the largest, reddest, and best-tasting tomatoes in the whole county. This scientific process is known as
    a. survival of the fittest. b. resistance.
    c. (selective breeding.) d. variation.

18. Fishermen once tried to eliminate starfish by cutting them into pieces and throwing them back into the sea. This, however, did not solve their problem. It only made it worse. What characteristic of starfish (an echinoderm) explains why the situation worsened?

    Regeneration; each piece underwent regeneration and became a complete individual, which greatly increased the number of starfish.

Name ______________ Date ________ Class ________

SourceBook Assessment, continued

19. If you were digging deep in your backyard and found a piece of stone with the imprint of a starfish, could you find a scientific explanation for how it got there? (Your neighbor burying it there is not a scientific explanation.)

Answers may vary, but should involve a description of the Earth's changes over time. For example, one answer may be that the backyard was covered by ocean at one time.

20. Why are albino Bengal tigers more scarce than non-albino Bengal tigers? (Assume that albinism is just as common as non-albinism in Bengal tigers.)

Answers may vary, but the following idea should be included: The white color makes albino Bengal tigers less able to camouflage themselves in their environment; therefore, they are more apt to become prey (of hunters) or, on the other hand, to be less successful in stalking their prey.

21. The bones of a person's arm and the bones of a whale's flipper are arranged similarly, but they differ in size and shape. What can you infer from this observation?

Students may infer that the whale and the human may have had a common ancestor.

SourceBook

Name ______________ Date ________ Class ________

SourceBook Assessment, continued

22. Only mammals are warmblooded.
  a. true b. (false)

23. Birds have hollow bones that better enable them to fly.
  a. (true) b. false

24. From what we have learned about the history of the Earth, mammals were among the first organisms to appear on Earth.
  a. true b. (false)

25. All organisms in the kingdom Protista are multicellular.
  a. true b. (false)